NATHANAEL-ISRAEL ISRAEL, PhD

Science180 Accurate Scientific Proof of God

OTHER BOOKS BY NATHANAEL-ISRAEL ISRAEL

Get them at your local bookstore, or online (e.g. on Amazon, Science180.com/books)

Turbulent Origin of the Universe
There is Only One Scientific, Simple, Safe, Trustworthy, Unexpensive, Brave, Practical, Nonconformist, Universal, Verifiable Formula that Accurately Decodes the Universe Formation … But You Are Not Using It

Reconciling Science and Creation Accurately
What Science Accurately Teaches about Creation and God's Existence that Atheists, Freethinkers, and even Most Christians Ignore … And How to Demonstrate it Without Taking Sides Between Rationality and Faith

Turbulent Origin of Chemical Particles
Why You Don't Have to Embrace Evolution, Big Bang, or Deny God to Scientifically Prove the Formation of All Chemical Particles

Origin of the Spiritual World
Top Secrets about the Origin of Everything in the Universe that Some Elites Have Hidden from You for Thousands of Years

From Science to Bible's Conclusions
How Decoding the Universe-Origin by Properly Revisiting Scientific Data—That Top Scientists Collected but Wrongly Analyzed—Bizarrely led to the 3500 Years Old Biblical Account of Creation

Turbulent Origin of Life
Why You Don't Have to Embrace Evolutionism or Check Your Brain at the Door in the Name of Faith or Science to Accurately Decrypt the Origin of Life Using the Historic Formula of the Universe Formation

How God Created Baby Universe
What Children Must Scientifically Learn Early about the Universe Formation to Avoid Dangerously Abandoning God Later in Life Just Like Most College Students Who Embrace Evolution and Big Bang That Deny Biblical Creation

How Baby Universe Was Born
How to Scientifically Talk to Children about the Universe Formation and They will Know Forever How to Correctly Test the Intersection of Science and Faith

Mathematical Proof of God's Existence at the Intersection of Science and Faith.
The Scientifically Verifiable Cosmological Theory that Challenges the Big Bang Theory at the Crossroads of Reason and Religion THEY Want You to Ignore

More books written by Nathanael-Israel Israel can be found at Israel120.com/books

NATHANAEL-ISRAEL ISRAEL, PhD
Founder of Science180, www.Science180.com
Father of Science180 Cosmology
Discoverer of the Universe-Origin Formula
www.Israel120.com

Science180 Accurate Scientific Proof of God

Can We Scientifically Explain the Formation of the Universe Through Natural Processes Without Evoking Evolution and Big Bang?

$$T = \frac{D}{Ve} + \frac{2R\pi}{Vo}$$

Science180
Augusta
United States of America
www.Science180Publishing.com

Copyright © 2025 by Nathanael-Israel Israel
Visit the author's website at Israel120.com

Science180 Accurate Scientific Proof of God
Can We Scientifically Explain the Formation of the Universe Through Natural Processes
Without Evoking Evolution and Big Bang?

First edition: October 2025

Published by Science180
Augusta (USA)
www.Science180Publishing.com

Book Cover and Illustrations by Nathanael-Israel Israel

ISBN: 979-8-9932150-0-6

Library of Congress Control Number: 2025920900

More books by the same author can be found at Israel120.com and Science180.com

For information about special discounts available for bulk purchases, please visit Science180.com/discount for more details.

Science180 can bring authors including Dr. Nathanael-Israel Israel to your live or recorded events. For more information or to book an event, please visit Science180.com/speaking

For any questions, please visit Science180.com/contact

To publish your book(s) with Science180, go to Science180Publishing.com

To interview the author of this book, visit Israel120.com/interview
To donate, please visit Israel120.com/donate or Science180.com/donate.

Printed in the United States of America.

CONTENT

1. CAN YOU SCIENTIFICALLY PROVE THE EXISTENCE OF GOD TO ATHEISTS AND ALL OTHER FREETHINKERS WITHOUT UPSETTING THE BELIEVERS?

- If so, can science prove it in a simple language that all can understand plainly?
- If so, how can this rational demonstration be done quickly without fail?
- If not, did science disprove God?
- Is there a simple scientific formula accurate enough to clearly point us toward the God who created the universe, if he exists?
- Can we really explain the formation of the universe through natural processes without invoking evolution and the Big Bang theory?
- Is there any God that smart atheists, agnostics, rationalists, and other freethinkers can actually like using reason and scientific facts?
- Do people, including most fervent creationists, believe lies about creation that need to be changed so atheists and all other freethinkers can enjoy the God they hate?
- Do freethinkers really hate God?
- Do many believers deny God more than they think, even when they consider unbelievers as the only God doubters?
- Is there any specific thing smart people can scientifically do to better know or test the inerrancy of the creation story while keeping their reason intact and free?
- Are believers really scientifically proving the existence of God to rationalists and freethinkers like they say?
- Is there any real science behind the creation narratives that can practically and scientifically benefit humankind?
- Can anyone scientifically prove the inerrancy of any creation narrative without quoting the Bible, the Quran, any other religious book, or any story related to dinosaurs and Noah's ark that annoys the secular rationalists?

- Can smart freethinkers actually end up liking the God they can't rationally prove exists?
- Is there any scientific story that atheists, evolutionists, agnostics, rationalists, humanists, and all other freethinkers can properly scrutinize and automatically rethink their argument against the existence of the God they said they know doesn't exist or is dead?
- What if science teaches something crucial about creation and the existence of God that most believers, atheists, and freethinkers ignore?

| Can smart freethinkers actually end up liking the God they can't rationally prove exists? |

- Can we scientifically start from nothing or near nothing and really get to the complex world we have today?
- What if the Abrahamic religions scientifically teach something about creation that believers refuse to demonstrate because they don't know, yet they believe it and want rationalists and freethinkers to do the same in the name of faith that is said to be at war with science?
- Is it a waste of time to attempt to rationally test all creation stories (including the Biblical creation and the Islamic creation) using science or historical investigation? Is that even scientifically possible?
- What if I, Nathanael-Israel Israel, can scientifically help you get closer to the real answer to those thorny questions without forcing you to pause or reject your reason or mind, doubt, belief, or faith?

By the time you finish reading this book, whether you are a believer or a doubter or any kind of freethinker, you will answer most of your critical questions about God's existence confidently. If you are interested in the most accurate scientific test of the existence of God but did not know how to prove or scrutinize it, get ready to rationally answer that question once and for all.

Indeed, the debate about the existence of God has preoccupied humankind for thousands of years. Pundits and laypeople from various backgrounds have reflected on the topic using methods classified as empirical, logical, metaphysical, scientific, and subjective. For instance, some people have deployed scientific methods to try to disprove or prove God. Some people think that there is no God. For others, gravity proved that God does not exist. According to the late British physicist Stephen Hawking, God did not create the universe: gravity did[1]. Some people, like Friedrich Nietzsche, said that *"God is dead"*.

Like some atheists brought it up, "Can you know that something exists that you cannot understand?" Can you know that God exists if you cannot understand him? Maybe, like most atheists, because you don't understand God, you focus your attention on understanding others and serving them. Perhaps, because you cannot prove God scientifically, you refuse to believe in him, or maybe you believe in him or in the existence of a higher power despite the lack of scientific proof.

CHAPTER 1: CAN YOU SCIENTIFICALLY PROVE THE EXISTENCE OF GOD TO ATHEISTS AND ALL FREETHINKERS WITHOUT UPSETTING THE BELIEVERS?

Perhaps you disagree with some creationists who think that asking God to prove himself is dishonest and is a mistake. If you are a rationalist, you may view this kind of statement as unconvincing. I understand your pain when some believers say that asking for proof of God's existence is not a scientific question.

Maybe you are like those who think that, if God exists and wants to convince you about himself, he will know exactly what he would have to do to convince you. Therefore, because he has not done exactly that to you yet, according to your standards, you think or know he does not exist. Maybe you support the idea that if God really exists, and if he is the most logical being, he should have known that the smartest and wisest thing ever for him to do is to prove himself to everybody plainly beyond the shadow of a doubt, not through scriptures, parables, fake stories, or religious nonsenses, but through factual science, not based on what you perceive as a flawed cosmological and moral argument. In simple terms, you just think that there is no place for God in science.

Maybe you believe that, not only is there no God, but also that there is nothing after death, and that after death, your body will just decay, and that's it. Perhaps you think that believing in God without demonstrating it is an intellectual laziness. In other words, you think that God is the laziest explanation of the immense complexity in the whole universe. Who knows if you consider believers in God as idiots, ignorant, scientifically illiterate people who don't deserve to teach science, but who need to learn real science from rationalists like you?

If you are like Dan Baker, the fundamentalist evangelist preacher who dumped Christianity to become an atheist, you may say that, "if nothing comes from nothing, God cannot exist. Because God is nothing, you think that God must have come from something; you think he is not God. Moreover, like Dan Baker and many other atheists, you may think that God does not exist because he cannot be omniscient, cannot change what he would do tomorrow, and still be free at the same time. In other words, like many atheists, you think that God cannot perfectly know the future, even his future actions, and be free at the same time. For you, a nonbeliever, if God cannot change what he is going to do tomorrow, you think that it puts some limits on his freedom and on what he can do, you think like Dan Baker that he is not free and omnipotent.

Peradventure, using the statement of Apostle Paul (author of many books in the Bible) who wrote that "God is not the author of confusion," you believe that there is no God because, for you, no book has caused as much confusion as the Bible. If not, you may add, why is agreement among believers lacking? You may even believe that the Bible is a collection of fables or nonsense. Or you think that if complexity requires a design, the mind of the complexity in the world must also require a designer; therefore, you think there is no God, for the mind he may have used to create a complex world requires a designer who is not him. Like countless atheists, you may support the resurrection of Jesus, which most Christians believe is God, while most Jews (except the Messianic Jews) deny he is even the son of God, to even dare to be God himself, is the worst example of evidence or argument for the reliability of the

SCIENCE180: BREAKING COSMOLOGICAL AND TRADITIONAL NONSENSES ABOUT GOD'S EXISTENCE

Bible. Maybe you don't believe in the existence of God because you think that the existence of God cannot be scientifically proven. Just like some agnostics or atheists, you think that science and religion cannot coexist. Before I move any further, let me present some statistics about atheism in the USA.

Indeed, a poll conducted by the Pew Research Center in 2023 showed that 4% of Americans self-identified as atheists[2]. This is about a 1% increase from the same poll[3] conducted by the same organization, Pew Research Center, in 2014. In the aforementioned survey conducted by the Pew Research Center in 2023, about 23% of atheists believe in the existence of a higher power, but not a god. According to a 2023 poll conducted by PRRI (Public Religion Research Institute), about 5% of Americans identify as agnostics. In 2022, in a poll[3] conducted by Gallup, 17% of respondents said they "Do NOT believe in God?" In general, the number of people who do not believe in God is increasing. But can anyone blame or credit secular science for all of that?

In contrast, billions of people across the globe have turned to religion to answer the challenging questions of God's existence. Some believe in God but do not wholeheartedly accept the creation story. Some say God used evolution to create the universe. Others believe that God created the universe in 6 days, but they are unable to scientifically demonstrate it to the rationalists or to publish their findings in the most famous high-impact scientific journals in the secular world.

As of 2025, the world has thousands of religions, of which the 5 most dominant are Christianity, Judaism, Islam, Buddhism, and Hinduism. Most of these religions have a leader or a founder, and a set of rules or stories pointing at or implying the existence of a supreme being that some view as the Creator. For instance, as it is written in the Bible, Christians and Jews call God by many names, including Adonai, Emanuel, Yahweh, Yeshua HaMashiach, or Jesus-Christ (that most Jews deny is God). In contrast, Muslims believe that God is Allah and that the holiest book is the Quran. In contrast, Buddhists have no creator God, while Hindus have many! If we can expand the world's top religions to Animism, the names and number of the supreme beings perceived as God will be astronomically shocking.

But to be honest, can we really say that any religion in the world has properly proven scientifically, without a shadow of a doubt, that God exists? Is such a demonstration even possible?

If there is God, how can we scientifically know which one of the beings, idols or gods, worshipped by the world's religions is the true God? If no God or leader of the above-mentioned religions, ideologies, or philosophies is the true or correct God, how can we scientifically prove this using raw data we can verify and test? In other words, if God exists, how can anyone use pure science only to rationally prove it to rationalists and all other freethinkers who don't want to believe in anything unless rationally proven? If we can prove the existence of God scientifically, can we be sure that all believers will actually accept the scientific version of the story if it does not match what they think their religious books say?

If you are interested in impartially and scientifically answering any or all of the questions mentioned above, and even more, then you are in the right place.

This book will help you to:

- have peace of mind that all your God's existence is precisely answered
- improve your understanding of the scientific test of God's existence without needing to deal with complicated theories or religious dogmas
- protect yourself and dear ones against biases, crashes, and harms due to errors and misinterpretations, therefore significantly making your efforts to scientifically understand God's existence safer and long-lasting, using scientific facts and tested proofs
- give you a unique perspective of how to probe God's existence, with capabilities exceeding that of any other existing God's existence theory, therefore providing you with advanced tools to explain complicated tasks as you think about or work on the cosmos and its content
- scientifically teach you the accurate and effective test of God's existence so you can precisely know the proofs you have been looking for to finally solve the difficult God's existence questions

Although primarily written for all types of freethinkers (e.g. atheists, humanists, rationalists, agnostics, nontheists, nonreligious, skeptics, nonbelievers, religiously unaffiliated, spiritual-not-religious, ex-believers, doubters) who know, think, or believe that there is no rational justification for believing in God, or who are not sure whether there is a God or not, or who are not even interested in exploring that question, this book can help you even if you are believer. In short, regardless of your faith or level of doubt or worldview, you will find this amazing book very useful as long as you identify yourself with any of the following descriptions:

- You are an atheist who thinks that God does not exist
- You are a strong believer in God
- You think that God exists, but you don't believe in him
- You think that it is impossible, pointless, or useless to scientifically demonstrate God
- You believe that science can never prove God
- You think that creationists are stupid and irrational
- You know that, even if someone presents you with the proofs of God's existence, you will never believe in him
- You have no interest in questions related to God or any deity
- You are a believer who is doubting God or who doesn't know how to scientifically investigate God's existence
- You are a believer who dislikes nonbelievers asking for a sound scientific demonstration of the existence of God
- You are a former believer who renounced your faith because you concluded that God does not exist

- You believe that God is dead, just as Friedrich Nietzsche said
- You believe that God exists, but you don't know which God it is or how to scientifically check the gods across the world's religions to know which one created the universe
- You are just curious about what I have to say

God's existence questions come with many problems, pains, and expectations, according to what people desire. Because it is unwise to begin solving a problem without first understanding all its sides as well as possible, I will spend the next chapter critically reviewing the main problems associated with the existence of God. Then, I will tell you how and if I will handle all of those problems in this book. Afterward, I will submit my findings to your attention before elaborating on why I think my discovery scientifically tests the existence of God. Throughout this book, I will introduce you to valuable resources that can help you or your loved ones in the journey to scientifically understand the origin of the universe, so you can improve your life and make more profit in whatever you do with peace of mind.

I also have valuable resources that can benefit you at www.Science180.com and on my personal website, www.Israel120.com. On these sites, you can also reach me for questions, comments, ideas, suggestions for partnership including donations and how to invite me to your event or organization so I can speak to your people or train them (according to your needs) in what I do best: scientifically decode the origin of the universe, of chemicals, and of life, and scientifically testing the existence of God. In the next chapter, I will share a very important secret with you: how I killed the most crucial myths about the origin of the universe and God's existence.

2. HOW TO IMPARTIALLY INVESTIGATE GOD'S EXISTENCE PAINS AND EXPECTATIONS IN THE WORLD TO PROPERLY HANDLE FAITH AND DOUBT SCIENTIFICALLY

- What pains do you experience when trying to address God's existence problems?
- What risks do you take when trying to solve God's existence problems?
- What emotion or annoyance do you experience or encounter while trying to solve problems related to God?
- What are you hoping to gain from any effort related to testing or proving God's existence?
- What are you expecting from this book?

2.1. Undeniable God's existence problems that any serious demonstration must address

Without a doubt, although they may not admit it, people of all backgrounds and belief systems struggle with the scientific demonstration in favor of or against God's existence, at least in the eyes of their opponents.

I understand that everyone reading this book expects to gain something significant from it, and I can promise you that you will. I also know that everyone who is trying to scientifically test, prove, or disprove the existence of God meets some challenges.

The pain points of God's existence vary, whether you believe in God or not. Therefore, to be sure, I need to clarify what I know you will gain from this book by first reviewing what most people would expect when addressing problems of God's existence. For instance, what a freethinker would expect is not what a believer would. Those who believe in God disagree among themselves about their proofs, just as those who deny God evoke various reasons to justify their doubts.

From the far left side of atheism where you see people (like Richard Dawkins)

7

saying they know God does not exist, to the far right side of creationism where you see people (like Carl Jung) who don't need to believe in God, but who know that God exists, there are various answers and nuances to the question of God's existence. However, all of these people can be grouped into two categories: believers and non-believers; between them are variants I cannot detail here. However, I scientifically address all of their common denominator in this book. But what is that common denominator? Let's investigate it first!

Indeed, most creationist books are framed as defenses or attacks against anti-creationists. Likewise, most anti-creationist books are crafted as if the creationists are out of touch, irrational, insane, and unscientific enemies. The creationists and the anti-creationists do not really understand each other. They don't seem to be trying to see God's existence from the perspective of their opponents. In other words, they seem to want to convert their opponents to their viewpoints without listening carefully to them. Therefore, most creationists can give long, and sometimes inconsistent speeches, ignoring the science, as if reading Bible verses will miraculously cause the rational unbelievers to trash their secular theories and start singing Alleluia! Amen!

At their turn, the anti-creationists are quick to bring out their evolutionist and Big Bang theory demonstrations as if those arguments will sway or deter the staunch creationists to drop their faith or the God, which they are ready to take bullets for or die for, even if some of them will swiftly deny that God if real problems are dropped on their career, wallet, or secular education!

- Can the rationalists and all other freethinkers address the believers as if their faith is scientifically useless, while the creationists judge the rationalists as God-haters and dangerous, yet expect to find the support needed to advance science for the benefit of humankind?
- Can the creationists and the anti-creationists solve the problem of God's existence without empathically caring for the pains of their opponents and avoid falling too quickly into the trap of unconsciously taking sides in the battle between science and faith?

I understood that we cannot holistically solve God's existence by forcing all believers to drop their faith to please and kiss secular science, while asking the rationalists and all other freethinkers to deny science and embrace faith blindly as the solution to their scientific problems or to the problems in the world, which the atheists also use to deny God. I understand that, if you are an atheist, because you see disasters, suffering, evils, and pestilence almost everywhere on Earth, you do not think that an all-powerful, smart, and mighty God can exist.

- But, is using the major problems in the world the greatest scientific proof against the existence of God?
- Can we at least try to get some help from math to scientifically see if any scientific theory can accurately lead us to faith, or vice versa?
- Or can we rationally review the faith across the world to see if its content can lead us to any real science?

2. HOW TO IMPARTIALLY INVESTIGATE GOD'S EXISTENCE PAINS AND EXPECTATIONS IN THE WORLD TO PROPERLY HANDLE FAITH AND DOUBT SCIENTIFICALLY

- Can science lead to any religious truth about the origin of the universe, or can any religious story of creation have a scientific power that can help us test God face-to-face?
- Can we be nonjudgmental and tackle these critical issues, letting the data lead us to the correct scientific conclusion?

If you are reading this book, you are probably looking for indisputable proof to scientifically test, understand, or convince yourself or others about the existence of God or not. Maybe you are seeking how to use that insight to positively transform lives, develop yourself, and advance humankind. Perhaps you are an amateur or a professional who just wants to scientifically determine whether there is a God or deity who created the universe. Or you feel like the existing theories on God's existence are weak, and you know there may or must be a better scientific theory, but you just don't know where and how to figure it out without checking your brain, faith (or doubts) at the door! Who knows if you have been researching scientific proofs of God's existence, but, with the countless data and theories available on the internet or across the globe, you are not sure of the path that can scientifically lead to the truth? Perhaps you don't know how to scientifically test your convictions to gain peace of mind. It is possible that, in the process, you may have been confused about God's existence, and you are unable to clearly and accurately state how to test your conclusions or tendencies on this hot subject that some people say has eternal consequences that you ignore because, for you, it is irrational! Maybe you have wondered whether your belief, doubt, or disbelief about God's existence is 100% correct, and you want to scientifically ensure you are on the right track instead of taking the risk of holding onto a wrong assumption, philosophy, religion, or scientific theory until it is too late! Maybe, because you don't believe in any timeline after death, or because you don't even care about any deadline, you don't think that you have to even think about God.

Who knows if you are already very sure about your worldview of God's existence, but you just don't know how to scientifically challenge it using the scientific data and the right perspective that can rationally convince both believers and unbelievers? Maybe you are a rational decision-maker interested in a truly original rational argument of the existence of God (supported by scientific data). Or you just want a simple, yet accurate origin proof that can help you understand advanced mysteries about God's existence without taking sides between mind and faith, or between science and religion. I cannot rule out that you may have been overwhelmed by the amount of data and books available on God's existence and you want a theory that quickly answers all your critical origin-related questions in one place. If you have any of these problems, know that you are reading the right book to help you.

What I can say for sure is that most people want:

- a safe, secure, and hassle-free way to test, prove, and eventually understand God's existence holistically

- a God's existence theory that encompasses most of the data available on the cosmos and that increases your performance, confidence, and safety
- a proper understanding or test of God's existence to make you happier, to add meaning and satisfaction to your life, to make your job easier, and cause you and your audience to better respect you and listen to you
- a way to increase the number, abilities, skills, and productivity of yourself and maybe of your customers, seeking to scientifically understand the existence of God, while boosting your reputation, profit, and keeping your mind above all nonsense, whether they are religious or scientific

But the pursuit of the above expectations or the solutions to these common problems of God's existence comes with many pains for most people:

- Most books about God's existence available on the market are usually complicated, inappropriate, partial, and erroneous
- Understanding and navigating all of the questions related to God's existence is complex, painful, inconvenient, and irritating
- Many difficult, technical words get in the way of a straightforward experience of decoding God's existence
- With the abundance of religious stories and scientific theories about the origin of the universe, of chemicals, and of life, you can be easily overwhelmed with the number of options you have while trying to address the unknowns, including God that you cannot see with your naked eyes
- You may feel like what some people believe or deny is preventing them from unlocking the mysteries that should lead you or them to the reality about the nonsense surrounding God's existence
- You feel like, although a lot of data is collected on the universe, scientists and laypeople are failing to properly interpret them to test God's existence because the correct examination of these data requires a cross-disciplinary insight that most people lack

Although the problems I mentioned are universal, nuances exist, and to properly address the problems of God's existence, we cannot ignore or pretend to know the pains of the nonbelievers and of the believers. Therefore, at this point, before I go any further, wouldn't it be appropriate to listen to what freethinkers have to actually say against the existence of God? If you disagree, please go to section 2.3.

2.2. Atheist God's existence pains and expectations

If you are an atheist or any other kind of rationalist or freethinker,

- You may consider most religious people insane, crazy, and delusional
- You support critical thinking, and you oppose religious views in government and public policy
- You want to connect with others who have left belief behind

2. HOW TO IMPARTIALLY INVESTIGATE GOD'S EXISTENCE PAINS AND EXPECTATIONS IN THE WORLD TO PROPERLY HANDLE FAITH AND DOUBT SCIENTIFICALLY

- Your worldview is naturalistic, free of mystical or supernatural elements
- You oppose religious lobbying and promote a secular world
- You oppose laws that prohibit blasphemy
- You don't concern yourself with the existence or non-existence of a divinity as much as you want to care for people
- You love yourself and your fellow men instead of God or a god
- Perhaps religion has brought shame to you because you left your faith (apostasy), or you never believed in God or any divinity at all
- You believe that most public and ethical progress was brought about by people free from religion like you
- You are against church politicking, church campaigning, and governmental subsidies offered to churches
- You believe that all ideas and religions must be examined or scrutinized using science, critical thinking, or reason
- You probably don't believe in any world other than the natural world
- For you, what can really save humankind is critical thinking, not religion or faith
- You tentatively accept findings that are supported by data
- You believe that government and religion must be separated absolutely, and this causes you to oppose prayer and Bible instruction in public schools
- You accept that prayer cannot help you, but your help must come from within your innermost
- Perhaps you agree with the skeptic that, if believers can prove God's existence to you through rational arguments or empirical evidence, then you will be glad they show it to you
- Maybe, like some Positive Atheism advocates would say, *"If anyone thinks he has a truly original argument to present to you, you will do your best to give it a fair look."*

If you are an evolutionist, you may experience certain specific pains of God's existence :

- You believe that creationists only criticize evolution, but never offer a comprehensive scientific theory of creationism
- You think that science is the single method by which humankind can understand nature
- You don't want to engage in or encourage any religious interpretation of natural phenomena

- You believe that the Earth is very old, that God did not create life, that life did not appear suddenly, and that all organisms have changed since their beginning
- In your mind, the universe does not descend from a supreme being
- You reject Genesis literalism or the Biblical account of creation literalism
- You are doing all you can to deny God and encourage others to believe that there is no God
- You believe that creationism poses a threat to science education
- You think that ignorance (not creationism) is the alternative to evolution
- You oppose the teaching of religious views in science classes
- As the National Academy of Sciences stated, you unequivocally believe that "creationism has no place in any science curriculum at any level"
- You think that creationist arguments against evolution don't hold
- You deny all religious interpretations of natural things or phenomena
- You oppose the teaching of creation and of any arguments against evolution
- You believe that any explanation that is inconsistent with empirical evidence or tests must be removed from science
- You deny religious or ultimate explanations, but you accept only empirical evidence
- You may celebrate that the teaching of Biblical creation in the US schools is legally ruled unconstitutional and unscientific
- You believe that biology is founded on evolution
- You consider that creationism is trying to erode quality science education
- You believe that those who think that evolution should not be taught in school are wrong
- You are against teachers using the power of the state to teach religion in classrooms
- You believe that evolution is a "major unifying concept" in science, and you support it being emphasized in schools
- You believe that without evolution, the literacy level of students and professionals will be limited
- You believe that all scientific domains or disciplines show evidence that evolution has occurred and is still occurring
- You like that evolution has infiltrated all areas of biology education
- You want all the lobbying efforts of creationists to teach creationism as an alternative to evolution in science classrooms to fail

- You think that the teaching of creationism in public schools would advance the religious belief of the existence of a supernatural creator and impermissibly endorse religion
- You support the judges who upheld that a teacher's First Amendment right to free exercise of religion is not violated if a school district requires that evolution must be taught in classrooms
- You think that no teacher has a free speech to override a curriculum, forcing the teaching of evolution in public schools
- You think that creationism cannot be confirmed or denied by using the scientific method; therefore, it must be removed from the science curriculum
- You think that evolution is the only explanation for life on Earth

Therefore, if you are a rationalist or an atheist, you are irritated by those who think you are bitter against God. In other words, if you are a rationalist and an atheist, you are bothered by people who insist that there is a God, and who try to show you where in the Bible it says so. You are annoyed by people who ask to pray with you or who invite you to go to church with them. You are displeased by those who think that, because the universe is very complex, it cannot just exist but must have been created by a God who "just exists". If you live in the USA as an atheist, you are pained by people who insist that the US Constitution is based on the 10 Commandments. Perhaps, you hate those who tell you that you are going to hell because you are an atheist or any other kind of freethinker. In other words, people who think that you are willingly ignoring the truth in creationism discomfort you. Hence, you dislike people who want to talk to you about God, but who ignore your questions and doubts about a supreme being. You feel bad when people insist that the Bible is not a myth, but a true book. Worst, you can't stand people who try to prove the existence of a creator to you by referring to Noah's Flood. Those who go further to say they saw Noah's Ark on conservative news outlets or TV stations anger you a lot.

- Can we really say that God exists because we saw the proof of Noah's ark on the news? Come on! I hear you! I got your pains!
- Can creationists really prove God's existence to freethinkers by referring to dinosaur bones, of which nobody knows how these giant animals became extinct, while the Bible did not even say much about them?

You are pained by religious people who repeat over and over their same argument about God as if it will make it true in your head. Nothing pains you as people who insist that everything in the Bible must be taken literally. You feel hurt when people tell you that God is mysterious and that you cannot rationalize his ways or that you are unfit to understand them. You dislike those who think that you should not question the morality of God when you see big crimes in the world. You disgust people who cannot prove their religious belief, but who insist that they are self-proven. You don't like hearing that faith is the only logical answer to your questions about God. Hence, telling you that God is real, just as the air you cannot see is real, irritates you a lot. Yes, you are irritated when some believers list top scientists who believe in God, while they refuse to name smart atheists who deny God.

You don't want to ever befriend some people because they think you are a fool because you doubt God. Moreover, you don't like people telling you that you have to believe before you can understand the evidence of God's existence. This may explain why you are irritated when Christians insist that the Bible is completely true or when they suggest that excellent things in nature prove God's existence. Hence, you brush these arguments as baloney or stupid nonsense. You may find these arguments sufficient to fight governmental funds used to help Bible schools or to repair or maintain churches, which are exempt from taxation and some fees[5].

If you are an evolutionist teacher, some key pains of God's existence can be yours to bear:

- You feel like you face pressure to introduce creationist teachings (that you think is nonscience) into science classrooms
- You are pained by religious people who want to force you to believe in or teach nonscientific things
- You feel pressured to include Biblical creationism materials into your evolution teaching
- You are frustrated by those who think that Biblical creationism must be taught in public schools
- You are against laws and governmental actions that promote religion or interfere with free speech
- You are pained by all the efforts of creationists to challenge evolutionism
- You are frustrated at creationists persuading politicians, judges, and others that evolution is a "flawed, poorly supported fantasy"
- Unlike what antievolutionists confess and wish, you don't want any form of Biblical creationism to reopen science classrooms to discussions of God

And when you are an evolutionist lawyer, you may face other problems I cannot ignore in this review of God's existence problems:

- You are troubled that teachers who believe in evolution are sometimes pressured to present creationism or to conversely downplay evolution
- You believe that the teaching of creationism in public schools misinforms students about evolution and weakens the scientific method
- You think that teaching creationism as science may cause students to struggle with post-secondary scientific education
- You think that creationists are discrediting evolution
- You are pained that some school boards adopted policies that open the door to presenting creationist critiques of evolution in biology class

14

2. HOW TO IMPARTIALLY INVESTIGATE GOD'S EXISTENCE PAINS AND EXPECTATIONS IN THE WORLD TO PROPERLY HANDLE FAITH AND DOUBT SCIENTIFICALLY

- You are angry that some boards of education voted that school textbooks must contain disclaimers cautioning that evolution is just a theory that has gaps and problems
- You even think that it is unconstitutional to insert disclaimers about evolutionism in school textbooks
- You think that allowing teachers to teach Biblical creation in public schools is like injecting religious beliefs into the school curriculum

Maybe, like the British writer and inventor Arthur C. Clarke, you think that "*Religion is the most malevolent of all mind viruses.*" Who knows if, like the American theoretical physicist Stephen Weinberg, you think that "*Without religion, we'd have good people doing good things, and evil people doing evil things. But for good people to do evil things, that takes religion.*" Maybe you agree with Thomas Jefferson, the 3rd president of the USA, that "*If there be God, he must more approve of the homage of reason than that of blindfolded fear*". Perhaps, just like Friedrich Nietzsche, you think that "*In Christianity neither morality nor religion come into contact with reality at any point.*" I don't know if you support the German philosopher Friedrich Nietzsche who said that "*There is not enough love and kindness in the world to give any of it away to imaginary beings.*" I cannot say if you don't agree with the Arab philosopher Abu Ala Al-Maari according to which "*The world holds two classes of men - intelligent men without religion, and religious men without intelligence.*"

If you are an atheist who wants to borrow words from Isaac Asimov, you would say that "*Creationists make it sound like a 'theory' is something you dreamt up after being drunk all night*"? If you are a freethinker, you may agree with the American inventor and businessperson Thomas Alva Edison that: "*So far as religion of the day is concerned, it is a damned fake. Religion is all bunk.*" Or do we just agree with Thomas Alva Edison, who said: "*I do not believe any type of religion should ever be introduced into the public schools of the United States.*" What about the American scholar Henry Louis Mencken, who said: "*Religion is so absurd that it comes close to imbecility... Religion is fundamentally opposed to everything I hold in veneration--courage, clear thinking, honesty, fairness, and, above all, love of the truth?.*"

Maybe you agree with the American television screenwriter and producer Gene Roddenberry that "*Religions vary in their degree of idiocy, but I reject them all. For most people, religion is nothing more than a substitute for a malfunctioning brain.*" If you are an atheist, you may agree with the French military leader and emperor Napoleon Bonaparte, who once said, "*All religions have been made by men.*" Maybe you oppose religion because, just like the German philosopher and political theorist Karl Marx, "*Religion is the opium of the masses.*" Is "*Religion is excellent stuff for keeping common people quiet?*" I don't know what you think about that, but if you are like Napoleon Bonaparte, you will answer Yes to those exact words.

I can't bet on whether you don't agree with the American author and engineer Robert Heinlein that "*Religion is a crutch for people not strong enough to stand up to the unknown without help. But ... most people do have a religion and spend time and money on it and seem to derive considerable pleasure from fiddling with it.*" It is probable that, like the American lawyer Gerry Spencer, you would "*rather have a mind opened by wonder than one closed by belief.*"

Like or unlike what the British Romantic poet Percy Bysshe Shelley said, "*If God has spoken, why is the world not convinced?*", you are convinced of God's existence, or maybe you are not. Maybe you agree with her that "*It is easier to suppose that the universe has existed for all eternity than to conceive a being beyond its limits capable of creating it.*" I just don't know and cannot know where you stand with all these views or whether you embrace them all.

Can this prove what the Indian independence activist Mohandas Gandhi once said: "*I like your Christ, I do not like your Christians. Your Christians are so unlike your Christ.*" Maybe you deny God because, as the American comedian George Carlin said, you "*would never want to be a member of a group whose symbol was a guy nailed to two pieces of wood.*" If you are like the French writer, historian, and philosopher Francois-Marie Arouet (best known as Voltaire), you will say that "*Christianity is the most ridiculous, the most absurd and bloody religion that has ever infected the world.*" While you may agree with this statement, other people may try to hit you for your freedom of speech. Voltaire also said, "*God is a comedian playing to an audience too afraid to laugh.*" I don't know if you will laugh at this one, which Voltaire also said: "*England has forty-two religions and only two sauces.*"

Or, just as Friedrich Nietzsche, you defined faith as: "*Faith: not wanting to know what is true.*" For you, people of faith are just entertaining lies. Maybe you jokingly think God is somewhere in the sky that you can see when you get there; therefore, like the Soviet pilot and cosmonaut, the first human in space, said while in space, "*I don't see any god up here*". Maybe, you concur with the British musician John Lennon that "*God is a concept by which we measure our pain.*" Hence, for you, we cannot talk about God while ignoring the world's pains.

If you think like the American comedian and television host Bill Maher, you will say, "*God has an ego problem; why do we always have to worship him?*"

Perhaps you think that believers who pray are just wasting their time. Who knows if you won't agree with the Ancient Greek philosopher Epicurus that "*If the gods listened to the prayers of men, all humankind would quickly perish since they constantly pray for many evils to befall one another*".

If you can borrow the words of Mark Twain, you may wonder: "But who prays for Satan? Who … has had the common humanity to pray for the one sinner that needed it most?" What can you say about that? Maybe for you, God is a dictator, hence just like the Austrian-born German politician Adolf Hitler, you may confess: "*Hence today I believe I am acting in accordance with the will of the Almighty Creator*".

Maybe, like Epicurus, you wonder: "Is God willing to prevent evil, but not able? Then he is not omnipotent. Is he able, but not willing? Then he is malevolent. Is he both able and willing? Then whence cometh evil? Is he neither able nor willing? Then why call him God?" Perhaps, just like the French nobleman and politician Marquis De Sade, you support that "To judge from the notions expounded by theologians, one must conclude that God created most men simply with a view to crowding hell".

Who knows if you won't laugh at the story told by the South African churchman, politician, and Nobel prize winner Desmond Mpilo Tutu: *"There is a story, which is fairly well known, about when the missionaries came to Africa. They had the Bible, and we, the natives, had the land. They said, "Let us pray," and we dutifully shut our eyes. When we opened them, why, they now had the land, and we had the Bible."*

How about what the English evolutionary biologist Richard Dawkins supports: *"We are all atheists about most of the gods that societies have ever believed in. Some of us just go one god further."* Are we really all atheists? Maybe, if you don't believe in God, you agree with the German-born French philosopher Baron d'Holbach that *"If we go back to the beginning, we shall find that ignorance and fear created the gods; that fancy, enthusiasm, or deceit adorned or disfigured them; that weakness worships them; that credulity preserves them, and that custom, respect and tyranny support them in order to make the blindness of men serve its own interests"*.

Whether you agree or not with everything I reported so far, I must confess that, during my investigations of the atheist pains, I came across some of what is known as the most famous rational quotes, but which, for the sake of time and space, I cannot comment on here, but which I cannot ignore either. Therefore, so we all can understand what the atheists and most other freethinkers think about God's existence, let me finish this segment with other favorite quotes of rational people[5].

- *"I recall the story of the philosopher and the theologian. The two were engaged in disputation and the theologian used the old quip about a philosopher resembling a blind man, in a dark room, looking for a black cat—which wasn't there. "That may be," said the philosopher: "but a theologian would have found it."* (Julian Huxley, British evolutionary biologist and author)

- *"Religion has actually convinced people that there's an invisible man — living in the sky — who watches everything you do, every minute of every day. And the invisible man has a special list of ten things he does not want you to do. And if you do any of these ten things, he has a special place, full of fire and smoke and burning and torture and anguish, where he will send you to live and suffer and burn and choke and scream and cry forever and ever 'til the end of time! ... But He loves you."* (George Carlin)

- *"The God of the Old Testament is arguably the most unpleasant character in all fiction: jealous and proud of it; a petty, unjust, unforgiving control-freak; a vindictive, bloodthirsty ethnic cleanser; a misogynistic, homophobic, racist, infanticidal, pestilential, megalomaniacal, sadomasochistic, capriciously malevolent bully."* (Richard Dawkins)

- *"The study of theology, as it stands in Christian churches, is the study of nothing; it is founded on nothing; it rests on nothing; it proceeds by no authorities; it has no data; it can demonstrate nothing."* (Thomas Paine, American founding father, philosopher, and activist)

- *"We must question the story logic of having an all-knowing/ all-powerful God, who creates faulty humans, and then blames them for his own mistakes"*. (Gene Roddenberry, American TV screenwriter and producer)

- *"With or without religion, you would have good people doing good things and evil people doing evil things. But for good people to do evil things, that takes religion"*. (Steven Weinberg)

- *"As far as I can tell from studying the scriptures, all you do in heaven is pretty much just sit around all day and praise the Lord. I don't know about you, but I think that after the first, oh, I don't know, 50,000,000 years of that I'd start to get a little bored"*. (Rick Reynolds, American comedian)

- *"Religion does three things quite effectively: Divides people, Controls people, Deludes people"*. (Carlespie Mary Alice McKinney)

- *"Faith does not give you the answers, it just stops you from asking the questions"*. (Frater Ravus)

- *If God is the alpha and the omega, the beginning and the end, knows what has passed and what is to come, like it states in the Bible, why do people pray and think it will make any difference"*. (Mark Fairclough)

- *"We would be 1,500 years ahead if it hadn't been for the church dragging science back by its coattails and burning our best minds at the stake."* (Catherine Fahringer)

- *"A myth is a religion in which no one any longer believes"* (James K. Feibleman)

- *"Just in terms of allocation of time resources, religion is not very efficient. There's a lot more I could be doing on a Sunday morning"*. (Bill Gates)

- *"All national institutions of churches, whether Jewish, Christian or Turkish, appear to me no other than human inventions, set up to terrify and enslave mankind, and monopolize power and profit"*. (Thomas Paine)

2. HOW TO IMPARTIALLY INVESTIGATE GOD'S EXISTENCE PAINS AND EXPECTATIONS IN THE WORLD TO PROPERLY HANDLE FAITH AND DOUBT SCIENTIFICALLY

- *"You can fool all of the people some of the time, and some of the people all of the time, but you cannot fool all of the people all of the time."* (Abraham Lincoln)

- *"Those to whom his word was revealed were always alone in some remote place, like Moses. There wasn't anyone else around when Mohammed got the word either. Mormon Joseph Smith and Christian Scientist, Mary Baker Eddy, had exclusive audiences with God. We have to trust them as reporters---and you know how reporters are. They'll do anything for a story."* (Andy Rooney)

- *"Truth does not demand belief. Scientists do not join hands every Sunday, singing, "Yes, gravity is real! I will have faith! I will be strong! I believe in my heart that what goes up, up, up must come down, down, down. Amen!" If they did, we would think they were pretty insecure about it".* (Dan Baker)

- *"A man is accepted into a church for what he believes and he is turned out for what he knows."* (Samuel Clemens)

- *"This would be the best of all possible worlds if there were no religion in it."* (John Adams)

- *"Lighthouses are more useful than Churches."* (Benjamin Franklin)

- *"Religion can never reform mankind, because Religion is slavery."* (Robert G. Ingersoll)

- *" Why is it often that, when someone points out the lies of another person, the person who lied gets angry, rather than abashed. Never is this more apparent than when the lie pointed out is a lie to oneself, and the anger never more fierce than when the lie is one of religion."* (G. Scott Wells)

- *"If you could reason with religious people, there would be no religious people"* (Gregory House)

- *"The family that prays together...is brainwashing their children."* (Albert Einstein)

- *"When you understand why you don't believe in other people's gods, you will understand why I don't believe in yours."* (Albert Einstein)

- *"I cannot imagine a God who rewards and punishes the objects of his creation, whose purposes are modeled after our own -- a God, in short, who is but a reflection of human*

SCIENCE180: EASILY UNDERSTAND COMPLEX GOD'S EXISTENCE EQUATIONS IN MINUTES *frailty. Neither can I believe that the individual survives the death of his body, although feeble souls harbor such thoughts through fear or ridiculous egotisms."* (Albert Einstein)

- *"The word God is for me nothing more than the expression and product of human weaknesses, the Bible a collection of honourable, but still primitive legends which are nevertheless pretty childish. No interpretation no matter how subtle can change this."* (Albert Einstein)

- *"It seems to me that the idea of a personal God is an anthropological concept which I cannot take seriously. I also cannot imagine some will or goal outside the human sphere... Science has been charged with undermining morality, but the charge is unjust. A man's ethical behavior should be based effectually on sympathy, education, and social ties and needs; no religious basis is necessary. Man would indeed be in a poor way if he had to be restrained by fear of punishment and hope of reward after death."* (Albert Einstein)

- *"If we are to believe the Bible, God created a flood that nearly wiped out all of mankind, taking genocide to epic proportions. Why then is God worshiped and Hitler, who committed genocide on a smaller scale, labeled evil for similar acts?"* (Scott Williams)

Most of the things I mentioned in this segment about atheist viewpoints are dear to most rationalists or freethinkers; they are real issues or concerns for most educated people who deny God, which creationists may not convince without handling properly. Yet, most creationists don't want to take these issues seriously. As I came to the end of this segment, my question to creationists is, "Can you ignore all of these quotes and still prove to people who disagree with you that there is God?"

My question to creationists is "Can you ignore all of these quotes and still prove to people who disagree with you that there is God?"

With all these problems they face, which I cannot ignore, freethinkers (e.g., atheists, humanists, evolutionists, agnostics) have specific expectations when it comes to proofs against or in favor of the existence of God and their own existence.

For instance, if you are an atheist, the understanding of how to fulfill your life through knowing yourself and your fellows may preoccupy you more than a proof of the existence of God. You hope for the world to be a secular place where public rules, scientific investigation, and education are free from religious influence and grounded in rigorous reasoning, rationality, and evidence. You wish that people who defy religious beliefs could freely speak, associate, and participate in public life. You support efforts to end physical pain and help people navigate life's struggles and trials. You advocate for religious people to deny God, gods, and all other forms of supernatural elements. You want city and school boards to stop public prayer. You don't want to see any nativity scene on public properties. You are against the censorship of freethought displays and the distribution of religious literature. You think that all privileges to religious groups must be abolished. You support that the government must be freed from all religious influence. And finally, but not least, you champion that all public policy must be based on reason and evidence.

If you are a humanist and I ask you about your expectations of God's existence, you may answer that you delight in forming your opinions based on reasoning, without being influenced by tradition, authority, or any established belief. You believe that the real meaning of life can be understood only by exploring human potential rather than the supernatural. You believe that only human thought, compassion, and the scientific method can solve the world's problems. If you are a humanistic Jew, for instance, although you consider the Torah (or the Old Testament of the Bible) as a valuable source of learning and inspiration, you don't mechanically accept it as authoritative. You enjoy reading and distributing critiques of apologetics that aim at persuading believers that apologetic arguments are flawed.

If you are an evolutionist, your God's existence expectations may be around the following:

- In the name of freedom of religion, you advocate that the government must keep its nose out of religion
- School boards should support teachers to emphasize evolution (not creationism) in curricula taught in public schools
- Science textbooks must highlight evolution but ban creationism altogether
- Just like the National Academy of Sciences of the USA said, you are against including creationist explanations in the processes of science because you think they cannot be tested, modified, or rejected by scientific means
- You agree with the National Academies of Science of the USA that "*Science teachers should neither advocate any religious interpretation of nature nor assert that religious interpretations of nature are not possible*". This also gives me hope that you will scrutinize this book thoroughly.
- You think that, unlike creationism, evolutionism does not assume either the absence or the presence of a creator, meaning that you are not completely against a possible scientific explanation of a Creator.

> Can creationists expect freethinkers to ignore all of their God's existence pains and expectations and just believe in a God they cannot scientifically argue with, touch, feel, or see with the naked eye, or with the most sophisticated electronic microscope, or the most advanced telescopes?

Can creationists expect freethinkers to ignore all of God's existence pains and expectations and just believe in a God they cannot scientifically argue with, touch, feel, or see with the naked eye, or with the most sophisticated electronic microscope, or the most advanced telescopes? Why don't creationists take the issues brought up by freethinkers seriously—or have they actually been doing that?

If you are an atheist trying to approach God-related issues only through reason that most believers want you to ignore, please know that I understand your frustration. I get it. I feel your pain when you deal with people who refuse to use their minds and focus only on faith that you or they cannot scientifically demonstrate! Part of this argument makes me glad because I know you will give my scientific test of God's existence a fair look. Hence, I am doing my best to address your very important questions with the best scientific care I can provide. Please keep reading, for I have something very important to show you in this book that will blow your mind scientifically!

If you are a believer who doesn't know how to appeal to unbelievers to scientifically demonstrate creation other than quoting the Genesis story, you may already be feeling mad at me for starting this chapter with atheist pains. But remember this book is primarily for very smart atheists, agnostics, and all other intelligent freethinkers who don't like the pseudoscience that some irrational believers point at them to argue the science behind the origin of the universe or the existence of God. Therefore, if you (a believer) do not like what I am writing so far, or if you think I should ignore the arguments of the anti-creationists, you may first check out my book called "*Reconciling Creation and Bible* Accurately," or you can just flip the pages to the segment I wrote about creationist pains. But if you can read this book until the end, you will better understand why most believers have been unable to use science to convince the staunchest, educated freethinkers, including the atheists and agnostics. If you disagree with me, please bear with me a little longer, and you will understand how some people are ignoring the plank in their own eye while they carefully look at the speck of sawdust in the eye of their "opponents" that they think (all of them) will go to hell, which means nothing to the unbelievers other than irritating them and making them ignore religion. Or is that statement simply false? If not, how can believers, even creationists, remove the speck in the unbelievers' eyes if, most of the time, there is a plank in the believers' own eyes as far as the real demonstration of the origin of the universe and God's existence is concerned? Therefore, until most believers honestly take out the plank in their creationist theories, they

> Until some believers honestly take out the plank in their creationist theories, they cannot clearly see to remove the speck from the eye of their rationalist opponents so the evidence or the test of God's existence can be demonstrated through the lens of pure science excluding faith—or isn't that possible?

cannot clearly see to remove the speck from the eye of their rationalist opponents, so the evidence or the test of God's existence can be demonstrated through the lens of pure science, excluding faith—or isn't that possible? I have come to realize that the difference between some creationist theories and some secular theories evoked by some crazy people is actually almost nonexistent. If not, how can a Biblical creationist believe that God took billions of years to create the universe, which the Bible says occurred in 6 days? How can we say that God, who is very intelligent, created a world that we cannot apprehend rationally using the mind that Christians say God gave us? Isn't that bizarre?

- Why don't creationist scientists take the criticisms of freethinkers and rationalists seriously?
- But what if Christians are not as irrational as most secular rationalists and freethinkers think they are?

Before I explain the problems and expectations I want to focus on in this book, don't you think it would be fair for us to also listen to the pains and expectations of believers regarding their perspective on God's existence? If you agree with me, please read the next section. If you disagree, please move to the next chapter. I know your pains, my friend.

2.3. Creationist God's existence, pains, and expectations

Do all creationists share any pain in common?

Although they believe or claim to believe in God, creationists have various views, some of which are not really distinguishable from those of the unbelievers. Space and time do not allow me to delve into all those differences here. Creationists often face resistance and misunderstanding from unbelievers, as they often fail to understand one another's problems and needs, leading to frustration when questions about God's existence arise. In general, if you are a creationist, the following can define you:

- You believe that God is the Creator of the universe and of all forms of life–whether you believe that God did it by himself quickly or through evolution, I don't want to debate that yet at this point, but later in this book!

- You know that science is not the only way that humankind can comprehend the natural world

- You can't understand why some unbelievers think that creationism poses a threat to science education

> How can a Biblical creationist believe that God took billions of years to create the universe, which the Bible says occurred in 6 days? How can we say that God who is very intelligent created a world that we cannot apprehend rationally using the mind that Christians say God gave us? Isn't that bizarre?

- You support the teaching of religious views in science classes
- You wish that a form of creationism would reopen science classrooms to discussions of God

- You may be pained that the teaching of creation (particularly Biblical creationism) in the US public schools is legally ruled unconstitutional and unscientific
- You support electing people who can cut funding to science, for you see science as against God
- You think that teachers can teach religion in classrooms
- You are struggling to convert unbelievers into accepting the creation narrative you believe in
- If you are a Christian, you believe that Christianity is the only true religion, but if you are a Muslim, you believe that Islam is the only true religion
- Depending on your religion, you defend that true science involving creation completely agrees with the historical explanations in the Bible or the Quran
- If you are a Christian, you believe that the Bible is the only absolute source of knowledge, but if you are a Muslim, you believe that the Quran is the only absolute source of knowledge. Later, I will also scientifically test if any or which of these 2 books correctly matches the science that I used to test God's existence
- You are concerned that some people think that there is a conflict between faith and science. In other words, you don't think that science conflicts with belief in God, and you do not see science as a threat that undermines faith
- You long to promote a healthy relationship and harmony between faith and the sciences
- You reject that religion and science exclude each other
- You are frustrated because the unbelievers you want to convert don't understand your proofs of the existence of God that you dearly believe are true, even if you cannot scientifically demonstrate them according to the standards of your opponents
- Rationalists, humanists, and atheists annoy you because they think the creation story of your religion is irrational
- You sometimes fail to properly answer the critical questions of some rationalists, who don't want to accept anything they think is illogical
- Perhaps, because you could not explain or understand the mystery of creation, you ended up embracing or creating wrong theories that go against the literal truth in the Bible or in the book of your religion
- Maybe, unable to discover the major flaws in your own ways of approaching God's existence scientifically, you attack or dislike people who want you to precisely prove your faith with raw empirical data
- You are pained that children and youth are regularly confronted by worldviews and moral standards that oppose God and cause people to question Biblical values, faith, or the doctrine of your religion

2. HOW TO IMPARTIALLY INVESTIGATE GOD'S EXISTENCE PAINS AND EXPECTATIONS IN THE WORLD TO PROPERLY HANDLE FAITH AND DOUBT SCIENTIFICALLY

- You are unhappy when secular people complain that creationist writings are scientifically incorrect
- You think that science conflicts with faith, or you believe that science is a threat that undermines faith
- You hate secular criticisms of creation
- You are pained that some teachers were punished for supposedly teaching Christian beliefs or the beliefs of your religion in science class in public schools

Because of what creationists believe about the existence of God, they expect certain things in life and from others. For instance, if you are a believer,

- You wish school boards, policymakers, and administrators would support and allow the teaching of some forms of creation in public schools
- You may long to see the truth about God properly embodied in doctrines and regulations, even political laws
- You want to educate and share with the public messages about creation astronomy
- You support distributing literature (magazines, books, tracts) about creationist articles
- You want to explore the relationship between science and the Bible
- You want to answer questions surrounding the book of Genesis, which you consider the most-attacked book of the Bible

Despite creationist complaints and expectations, as of 2025, anti-creationist perspectives expressed through evolutionism and the Big Bang theory are the dominant arguments taught in most public schools across the globe to deny the existence of God. It is only in private schools that creationist viewpoints are freely taught. In the USA, for instance, only Bible and Christian schools teach creationism to their students. In some Muslim countries where Islam is adopted as the national religion, the Islamic view of creationism (which is different from the Judeo-Christian narrative) is taught—I will come back to the difference between these 2 creationist narratives. But can we confidently say that the adoption of anti-creationist views in secular education implies that their perspective of God's existence is true?

Although most creationists believe that God created the universe, can we really say that they know how to scientifically demonstrate it? Or have they actually done it?

- Can the freethinkers expect the creationists to ignore their God's existence problems and just yield to rationalist criticisms? In the name of science and fairness, can we just dismiss the creationist pains concerning God's existence?

- Although most creationists believe that God created the universe, can we really say that they know how to scientifically demonstrate it? Or have they actually done it?
- Is it by dishonesty, ignorance, or incompetence that most of them will say they did?
- Should creationists be required to scientifically prove what they believe in before their belief is found or viewed as authentic? Does anyone even need a human being to authenticate any faith?
- Do some creationists feel obliged to use pseudoscience to prove the existence of God instead of pure science because they are ignorant or irrational?

Whether you are a Young Earth creationist (someone who believes that the Earth is less than 10,000 years old), an Old Earth creationist (someone who believes that God created the universe over a process of millions of years), or an intelligent design proponent (someone who does not necessarily publicly claim God as the creator but as an intelligent designer), your God's existence, pains, and expectations are unique, although they may not sound valid in the eyes of the anti-creationists and even in the eyes of some fellow believers who view creation through a different lens. Therefore, for the sake of precision, I need to briefly address those 3 kinds of creationist pains:

- Young Earth creationist God's existence, pains, and expectations
- Old Earth creationist God's existence, pains, and expectations
- Intelligent design proponents' God's existence pains and expectations

Young Earth creationist God's existence, pains, and expectations
If you are a Young Earth creationist:

- You believe that the God of the Bible created the universe and its content in six literal days as documented in the first chapter of the Biblical book of Genesis
- You don't believe that the universe or the Earth are billions of years old
- You believe that the creation story described in Genesis happened a few thousand years ago rather than millions of years ago
- You disagree with people, including some believers, who accept that God has used evolution to form the universe and life. In short, you believe that evolution is totally implausible
- You believe that the Bible offers a trustworthy, eyewitness description of the beginning of the universe and its content

- You believe that, unlike evolution, which is tearing faith and society, the Genesis creation account properly explains creation scientifically
- You believe that the days in the Genesis account of creation are not geologic ages, but are six twenty-four-hour consecutive days of creation
- You do not validate any interpretation of facts in any field that contradicts the Scriptures
- You believe that the origin account given in Genesis is a factual and reliable presentation of actual events concerning the history of the universe and life
- You believe that the Biblical statements of the origin of the universe and life are scientifically and historically accurate
- You support influencing politicians, judges, and others that evolution is a flawed, poorly supported imagination
- You believe that evolution is unscientific because it is not testable or falsifiable
- You are outraged that taxpayer dollars are used to fund evolution teachers, while the teaching of Biblical creation is rejected
- You think that evolution should not be taught in school at all
- You are convinced that forcing students to embrace evolutionary theory corrupts their minds and deprives them of the foundation for rational decision-making and belief in God
- You would like for evolutionism and the Big Bang theory to be torn apart completely
- Like Randall Hardy[6], you believe that atheists are just willfully suppressing the truth about the existence of God, but at the same time, you think that *"no amount of intellectual argument or proof will cause atheists to repent, because those* people have *already refused to yield the control of their lives to God."*
- You reject the claim that science has debunked God or that reality is limited to the physical world only
- You are dedicated to battling the evolutionary worldview and to promoting Biblical creation
- You believe that evolutionism is unquestionably brainwashing society
- You support the idea that science books must mention the "weaknesses" of evolution
- You support that school textbooks must contain disclaimers cautioning that evolution is just a theory
- You think that the teaching of evolution, but not of creationism, in public schools is a form of religious discrimination
- You think that evolution is not an explanation of life on Earth, nor a fundamental principle of life sciences

- You consider that discussions of evolution in classrooms interfere with children's free exercise of religion
- You think that the teaching of evolution must be forbidden in public schools, except if at least creationism is also taught alongside
- You wish your free speech right gave you the power to override a school curriculum, forcing the teaching of evolution in public schools
- You are pained that some federal judges have ruled that Biblical creationism is fundamentally a religious idea and that ordering its teaching in public schools would be unconstitutional
- You are pained by the widespread acceptance of evolution and its harm to millions who it deprived of a proper relationship with God, their Creator
- You believe that evolutionism has brought dangerous effects upon Western societies that embraced it and has robbed God of the glory for creation
- You feel like a ban on teaching Biblical creation in science classrooms is a violation of the First Amendment rights

Therefore, if you are a Young Earth creationist, you probably expect the following:
- Science textbooks should not highlight evolution, but Biblical creation
- You really wish policy makers and administrators would mandate the teaching of Biblical creation in public schools
- You wish that prayer can be recited in public schools
- You think that evolutionist teachers must be pressured to introduce creationist teachings (even if they think they are non-science) into science classrooms
- You want to expose the problems of evolutionary worldviews and defend the truthfulness of the Biblical account of creation
- You want to reveal the bankruptcy of evolution
- You want Christians to weaken and oppose evolutionism
- You want school boards to adopt policies that open doors to presenting creationist critiques of evolution in biology class

As you will see in the next segment, Old Earth creationists don't share most of the viewpoints of the Young Earth creationists, who, in return, may view some Old Earth creationists as unbelievers! By the time you finish reading this book, you will scientifically know who is wrong, who is right, or whether anyone is even wrong or right at all.

Old Earth creationist God's existence, pains, and expectations
If you are an Old Earth creationist, you will not agree with everything. Young Earth creationists say, but:

2. HOW TO IMPARTIALLY INVESTIGATE GOD'S EXISTENCE PAINS AND EXPECTATIONS IN THE WORLD TO PROPERLY HANDLE FAITH AND DOUBT SCIENTIFICALLY

- You believe in the allegorical interpretation of the Genesis account of creation
- You believe that the days in the Genesis account of creation are geologic ages, and are not six twenty-four-hour consecutive days
- If you are a Day-Age creationist, you believe that each day of creation was longer than 24 hours
- If you are an advocate of Gap creationism, you believe there is a gap of time between the days of creation or between the first verse and the second verse of the Genesis story in the Bible
- Perhaps you are a progressive creationist who believes that God created the universe and the forms of life gradually over hundreds of millions of years
- You don't believe that the Biblical account of creation is a strict chronological record of how God created the universe and its content, but you think that the Genesis story is a topical order of events
- If you are a Day-age creationist, you believe that the days of creation are not ordinary 24-hour days, but that they represent *"much longer periods of millions or billions of years"*
- You believe that all forms of life are related to a so-called great "Tree of Life"
- You believe that the Earth is billions of years old and that God created the universe and all its content over billions of years
- You believe that evolution is plausible
- Like Randall Hardy, you are *"convinced that it is a waste of time to attempt to 'prove the Bible is true' by means of science or historical investigation,* and you think that *"there is no need to prove the Bible to be true."*
- You accept the assumption that God used evolution to bring the universe and life into existence
- You may even believe that evolution replaced God
- You think that you can embrace an evolutionary worldview and still promote Biblical creation
- You don't support the idea that science books must necessarily mention the "weaknesses" of evolution
- For you, school textbooks do not have to contain disclaimers cautioning that evolution is just a theory
- You think it is OK to teach evolution but not creationism in public schools
- You don't think that it is important that students in public schools must be informed about the controversy concerning creationism and evolution
- You believe that evolution is a fundamental principle of life sciences

- You don't think that discussions of evolution in classrooms prohibit children's free exercise of religion, but you think that the teaching of evolution in public schools is good even when creationism is not taught alongside
- You think it is OK that a school curriculum forces creationists to teach evolution in public schools

Intelligent Design proponents' God's existence pains and expectations

What if you ignore the real God's existence problems of the intelligent design proponents? In fact, if you are an Intelligent Design proponent,

- You believe that certain features in nature can be explained only by an intelligent cause, not by natural selection
- You believe that the way nature is designed points to an Intelligent designer (which you don't publicly call God), not chance
- Unlike some creationists who usually refer to religious texts to make their case for God, you usually start with natural empirical results to draw inferences from the evidence, not to prove God, but an Intelligent designer
- You don't claim that science can identify whether the intelligent cause of the formation of the universe and life is supernatural
- Your goal is not to prove that God created the universe and life, as most creationists try to do
- You want the teaching of evolution to be discredited in public schools
- You believe that the concept of "irreducible complexity" favors Intelligent design and not evolution
- You think that evolution is not a scientific theory and that it lacks comprehensive scientific support
- You dislike people who think that Intelligent design is creationism lacking empirical support
- You are pained when creationists think that you have left the Bible out of your theory
- You dislike the US court ruling that Intelligent design is not science, and cannot disconnect from its religious antecedent
- You hate that the secular scientific community rejected the concept of irreducible complexity in peer-reviewed research articles
- You don't like it when people point at evolution through selection as a plausible explanation of the Intelligent design you see in nature
- You are frustrated that all creationist organizations don't embrace Intelligent design

- You disagree with people who think that Intelligent design is a vague concept because it is publicly divorced from Biblical creationism
- You are pained by people who think that your strategy of not naming the Intelligent designer as the Judeo-Christian God is unbiblical and a pure attempt to create a life-origin model acceptable by the secular community
- You are sad that some creationists think that the Intelligent design model will not work because, for them, it is not built using the Biblical method or Biblical creationism
- You dislike those who think that Intelligent design is meaningless because it does not publicly support Biblical creationism nor invite the church to return to its Biblical foundation
- You are frustrated when some believers think that Intelligent design is a "crude creationism" that has reduced the Creator God to a mere engineer
- You are sad when some people think that a goal of the Intelligent design strategy called "wedge strategy" is to produce a "theocratic state."

Therefore, if you are an Intelligent Design proponent, the following summarizes your expectations about God's existence:
- You really wish that policymakers, school boards, and administrators could mandate the teaching of Biblical creation in public schools as an alternative to evolution
- You think it is appropriate to require equal time for Biblical creation and evolution
- You wish that prayer could be recited in public schools
- You think that evolutionist teachers must be pressured to introduce creationist teachings into science classrooms
- You are eager to offer students a solid educational experience founded on Intelligent design evidence
- You wish that by not disclosing the Intelligent designer as the Christian God, you can strategically increase the acceptance of the Intelligent design theory, particularly in public schools

2.4. Nathanael-Israel Israel's pains and expectations concerning God's existence

Do you want to know my pains and expectations concerning God's existence so you know where I, Nathanael-Israel Israel, stand on this matter? Indeed, although I tried to summarize the main points of God's explanation and expectations about one. There were probably more things you disagreed with than you agreed with. But because I know that you understand that this book is for a large audience, you will bear with me, knowing that I am trying to appeal to everybody, even in the midst of the significant variations of belief and doubt that no one can ordinarily handle in a single book.

Just like you, I too have some specific pains and expectations that I would like to share with you. If I say I don't have any pain concerning God's existence, I lie, for during my investigation of how to prove the origin of the universe, of life, and of chemicals, which most people connect to God's existence or inexistence, I had serious pains, and I am still bearing some I cannot discount. If that is okay, please read on. Otherwise, please move to the next chapter.

As you have seen in the previous segments of this chapter, most people disagree a lot about things related to God. As you noticed, even the creationists disagree tremendously about the creation story, which they want the anti-creationists to embrace. At one point, I even wondered if, in the name of theology and what some may call revelation or even prophecy, some creationists even believe in God when they plainly deny what their religion book says about creation.

Among the anti-creationists, there are many schools of thought and it was sometimes hard for me to understand their nuances. Yet, they must be considered: freethinkers, atheists, humanists, rationalists, agnostics, nontheists, nonreligious, skeptics, nonbelievers, religiously unaffiliated, spiritual-not-religious, ex-believers, doubters, etc. I understood that:

- Both the pains and expectations of the creationists and the anti-creationists have a common denominator, yet these people fight as if they don't have any problems and as if their opponents are the problems
- Several people who have trained themselves in secular schools of thought are being left out partially because the conventional ways of sharing the creationist message are not fitting for their intellectual curiosity and methodology of thinking, seeking, finding, and receiving
- Several scientists are unable to discover the mistakes in their scientific interpretations of the world and how it was formed, and unfortunately, their scientific training is acting like a yoke that prevents them from believing in the God that they deny and defend cannot free them
- In addition to the unbelievers who reject the Bible and other religious books, many people who say they believe in God are misled, betrayed, and misguided to believe in scientific and religious lies, which apparently sound good and true to them, but which, in reality, are against the truth that even science can prove
- Several believers embrace wrong scientific theories (including wrong creationist theories) while kissing secular theories (e.g., the Big Bang theory and evolutionism) that they say oppose their religious book (e.g., the Bible), yet they accuse the nonbelievers of denying God, and in the end, they are like blind trying to lead other blind into a pit they don't know they ignore

- The problem is that human beings have been trying to acquire knowledge or interpret the invisible or the unknown using the wrong means and then apply what they think they know to try to achieve their goals, which don't always support reality, which is hard to distinguish from fiction in our modern digital world about to be dominated by artificial intelligence, which some people want to use to replace others who are thought not to be thinking as fast as robots, which we know can never surpass human beings' dynamic intelligence or unintelligence, if you want!

- Although some scientists didn't initially choose to deny God, the way data around them has been scientifically interpreted over time has caused them to rationally embrace theories opposing God, which even some believers are struggling to prove or disprove scientifically.

- Faith in God or in any deity or thing does not require physical proof. Still, some people's minds are so stuck in looking for proof for everything that they can't accept even things that are self-proven or self-explanatory or that may never be proven scientifically using modern rational standards.

> Every aspect of a test about the existence of God will never satisfy everybody, no matter the amount of proof presented.

- Therefore, if something is not done to help most people (of faith and of science) to know that their scientific data or spiritual dogma can correlate with a certain reality (that they reject or embrace unquestioningly in the name of science or faith), which some people are not looking for in their ideology or philosophy, it is near impossible for most people ever to come to know the real scientific test that can prove or disprove God's existence plainly in a way that even children ages 7-12 can rationally understand. In other words, something needs to be done to help people whose minds are stuck int.

Considering the aforementioned description of the problems, pains, and expectations associated with the existence of God, you agree with me that every aspect of a test about the existence of God will never satisfy everybody, no matter the amount of proof presented. There are deep expectations and pains associated with these problems that go beyond what science and even religion can ever solve. To put it another way, some expectations that people are looking for concerning the existence of God have nothing to do with science or faith at all. Hence, no human being can ever solve all problems connected to the existence of God. But can anyone ever solve all the problems of God's existence flawlessly without irritating people and even God himself if we can prove that He exists? So, what am I actually trying to accomplish in this book, then? What can I promise you you'll get for sure from this book? Read on, and you will see.

I understood that the real problem people have about God's existence is the scientific proof that can allow God's deniers to rationally review the creationist argument without leaving their brains at the door, while also allowing the creationists not to leave their faith at the door either. How can we do that?

- How can we accurately intercept faith and religion?
- Can we solve these problems without having science and faith come together?
- If you disagree, how then can anti-creationists ever convince the creationists without alienating them and considering them as irrational, and vice versa?
- Can we really say that all creationists are non-scientists?
- Can creationists and anti-creationists continue to make their opponents uncomfortable while still expecting them to help build the holistic support needed to advance their cause?

> Can anyone then ever solve all God's existence problems flawlessly without irritating people and even God himself if we can prove that He exist?

- In other words, can creationists keep embarrassing and fighting the anti-creationists, and vice versa, and expect that the support needed to advance science and faith be sustainably reached or built?
- Can't a dose of tolerance help to realistically engage the believers and unbelievers in the dialogue that may help to intercept or overlap science and religion wherever possible?
- Or can we just agree that science and religion, or reason and faith, are at war forever?

As you read this book, you will clearly know what you can save, do differently, become, and gain by paying attention to the historic secrets I had the honor to demystify. Whether you are a believer, an unbeliever, a freethinker, an administrator, a politician, a curriculum designer, a curriculum specialist, an education policymaker, a librarian, a school board member, a parent, a researcher, researcher, a student, a teacher, a member of the clergy, or a layperson, as long as you are really seeking to scientifically understand the rational test of the existence of God, "*Science180 Accurate Scientific Proof of God*" is the much-admired book written for great people just like you. As long as you are interested in the first and the only scientific book that talks to anti-creationists, evolutionists, Big Bang theory proponents, atheists, and all other freethinkers and rationalists about the universe, and they request to know more about the scientific proof of the God that they deny, then this book is for you. As long as you are really seeking to scrutinize the rationale of the existence of God scientifically, "*Science180 Accurate Scientific Proof of God*" is the book written for great people just like you. Don't wait any longer! As you read, you can also learn more at Science180.com/godproof.

3. HOW I DISCOVERED THE "UNIVERSE TURBULENT ORIGIN FORMULA" THAT YOU ARE NOT USING YET, BUT THAT IS THE ACCURATE KEY TO TEST GOD'S EXISTENCE SCIENTIFICALLY

How did I get involved in the research that led to this book? Because I knew many people would be interested in the answer to this question, I decided to address it first and introduce my background so you know how I got here.

In fact, for most of my life, up until 2024, I never planned to write this book. I was born and raised in a very poor country called the Benin Republic. If, like the 99.99% of the people I met in the USA these last 20 years, you are wondering where is Benin located, please know that Benin Republic is a small French-speaking country located between Togo, Niger, Burkina Faso, and Nigeria by the Gulf of Guinea (a large inlet of the Atlantic Ocean that borders on the southern coast of West Africa).

In Benin, I was told that agriculture is the engine of development. Therefore, I went to school all the way to university to become an agronomist, let's say an agriculture expert. I was first in my class when I graduated after more than 5 years of study at university.

After my graduation in Africa, life did not change much: the master of science holder in agronomy was still struggling like a sub-Saharan peasant, a poor farmer, who had not spent his last 20+ years on school benches, but under the Sun's zenith near the equator.

While in Africa, I previously learned that America, let's say the United States of America, is the greatest nation on earth, the land of the free and the home of the brave. Who doesn't want to be free from "poverty"? Therefore, without really questioning what freedom and happiness mean and whether or not that land that some movies painted as a heaven on earth (I don't know if my belief in those movies that I never watched myself was because the Africans who watched them did not know that movies are not reality), in pursuit of a better life in the country of Abraham Lincoln, I moved to the USA. But don't get me wrong: although China is trying very hard to take the lead worldwide, the USA is still the world's superpower, though not as it was portrayed decades ago. But will they stay on top forever, or what can they do to stay on top, or get back to the top if you think they are not there yet? By the time you finish reading this book, you will know if anything is needed from us, and what anyone and any nation can do to avoid falling. I love the USA, and I am glad I made it my home for more than 20 years. I know this book will also make the USA and many other nations better places to live if we follow some groundbreaking principles I will share very soon, based on my deep investigation of the origins of the universe, life, and chemistry.

After a few years in America, I attended school and earned my PhD in plant, insect, and microbial sciences at one of the best universities in the USA. By then, I was already a US citizen. During my graduation ceremony in America, I was the Doctoral Marshall. In other words, I finished my doctorate first among hundreds of PhD candidates in my class. I did not even graduate with my PhD before I got a job at a Fortune 500 company in the USA. I was earning a six-figure salary, with a company vehicle and other bonuses. From nowhere, by a concourse of circumstances I cannot explain in this book. Still, in another one coming up soon, because by itself, the description of those circumstances is more than a book, that job did not work out as originally planned. Despite some challenges I faced, I miraculously survived thanks to the wise precautions I had taken earlier. After a few months, I started having crazy ideas about things I had never considered before. Many questions crossed my mind:

- What else can I do to succeed in life— But was I really failing?
- How can a man who has "nobody" to help him make it in life, but did I really not have anyone to help me?
- What are the secrets of success?
- How can I be struggling in life while in Africa and even in the USA? I graduated first in my class, and I am a hardworking person who wants my dream to come true, but did it? Were my dreams actually unfulfilled?
- Why is being the best in school not a guarantee for a wealthy life?

I was asking some crazy questions that most people who have gone through bizarre experiences would ask. I got so many ideas that at one point, I could not resist writing them down. Therefore, I recorded these ideas on paper and electronic devices; some became the content of books I wrote later, and some remain unpublished but will be published very soon.

Initially, most of my ideas were about life and how to succeed in it, but as my thinking deepened, I also began asking questions about the universe itself. Hence, other ideas concerned the origin of the universe.

As I was transcribing notes on these "strange" ideas about 12 years ago, I decided to check what science had to say about them, particularly the universe's origins. During that search, I landed on NASA's (National Aeronautics and Space Administration) website, the American government agency that, among other things, conducts research on space, including celestial bodies in the universe.

CHAPTER 3. HOW I DISCOVERED THE "UNIVERSE TURBULENT ORIGIN FORMULA" THAT YOU ARE NOT USING YET, BUT THAT IS THE ACCURATE KEY TO TEST GOD'S EXISTENCE SCIENTIFICALLY

Once on that site, I saw for the first time some of the scientific variables commonly collected on celestial bodies. Each time I looked at a table or graph about these variables, a certain nonstandard trend and interpretation came to my mind. Whether it was the size, speed, density, movement, or other characteristics of the celestial bodies I was exploring in the Solar System, I was just seeing weird trends and understanding them clearly. I wondered if anyone had already seen and published what I was seeing, like trends in that data.

So I don't reinvent the wheel or waste my time thinking about basic stuff already well known, I decided to check the scientific journals to see what they said about those data. To my surprise, these journals did not answer the unusual questions I had, nor did they interpret the data as I felt they should. I quickly noticed that I have an uncommon story to investigate.

After years of investigation, I found myself with a large system of equations that I tried to understand. No one, nor any existing explanation, could satisfy me. In those days, life was very challenging, yet I have a clean record academically and societally:

- My academic grades (from high school through my PhD, which I earned with distinction in the USA) were strong.
- I never drink alcohol or any strong drink.
- I never smoked marijuana or any kind of cigarette.
- I don't have an immoral life.
- I don't have a criminal record.
- I have never had any issues with any governmental officers or men of order.
- My credit score in those days was great, and as of 2025, it is still great (in fact, except for my student loans (which are common in the USA and which the government lent me), I owe nothing to anyone).

So why then, someone like me, to whom people usually refer as the "handsome, intelligent, and smiley man," could not make it in the land of opportunities as I wish? If anybody tells you that, in those circumstances, life was easy, that person economizes the truth. Those with whom I shared my struggle could not believe me. On many occasions, I wondered if this was my destiny or whether something deep was happening beyond the visible. As you will see later in this book, and as I revealed in the 8 other books I published in 2025, something bigger than me was cooking, and I did not know it all at that time, but I knew I was into something bigger than my paycheck or my career. And when I finally accepted the sacrifice I had to make for a higher cause, you won't believe what happened next, which I will share with you.

Regardless of all this, I never lost hope and worked more than 15 hours every day, except on Saturdays. What do I do on Saturdays? Do I rest or do I farm in my yard to survive? Don't worry, you will know.

People who didn't know me called me by many names I cannot list here–or should I–, but in my upcoming autobiography, where I plan to blow the whole story and journey of the historic discovery I made.

Anyway, in the midst of all that, after carefully reading countless scientific articles on the topics that crossed my mind as possible explanations of the trends I saw in the aforementioned data, I realized that the solution of the systems of equations I was dealing with was dominated by the manifestation of turbulence, a problem that is considered as one of the unsolved mysteries in science and a mystery that has been puzzling scientists for more than 500 years, since the days of Leonardo da Vinci (1452–1519), the Italian painter, scientist, and engineer who made some of the earliest scientific observations of turbulence. By the way, turbulence is a term used to express the movement of fluids (e.g., water, air, any other gases or liquids, or materials that are not solid) under the influence of some instability. If you don't understand, don't worry; later, I will break it down for you, even if you are not a scientist or an expert in that field. And you don't need to understand turbulence to grasp the information I want to convey in this book.

Now that I knew that I was dealing with turbulence, I then asked myself: "What did the experts in turbulence say about turbulence?" I was shocked to discover that, for more than 500 years, little progress had been made in turbulence. During my search, I found that the first major conference on turbulence took place in Marseilles, France, in 1961, and the second one took place 50 years later, in 2011, in the same city. I read all their reports, but I discovered, as the experts in the field had confessed, that not much progress was actually made. I also noticed that the exports are very confused, and they designed countless complicated formulas filled with junk.

Knowing the questions the experts in turbulence have been asking, and knowing the trends I found in my original analysis of the data collected by great scientists on celestial bodies, I understood that the turbulence experts are stuck because the data and the precision they have been looking for cannot be found at the scale of their experiments in the laboratory and even in the field. I also realized that these data already exist and have been collected at the astronomical scale. Still, neither the turbulence experts who are looking for that data nor the physicists who collected it on celestial bodies knew that that data were actually about turbulence. In other words, I shockingly discovered that the turbulence holy grail that turbulence experts have been seeking for 500 years has been in the hands of physicists for almost as long. Still, the latter have been using "unapplicable" methods and asking the wrong questions (that some people may label as fictive) to unravel the secrets they contain. And because experts in most domains like to brag about what they think they know without trying to learn from what experts in other domains know, that data has not been properly used until I found it. And this problem is frequent in science but most scientists don't know: the key that one scientific domain is seeking can be available in another domain that the experts ignore, but those who can make the effort to explore things outside of their domains of expertise can better understand the big picture, which can lead to innovation.

What did I do when I found this turbulence holy grail? I crafted my original approach to investigate the data available on the essential bodies, integrating the force. The funny thing is that turbulence experts did not just lack the real data that should advance their field, but they also don't have a comprehensive and sound approach to study turbulence. Therefore, to study the celestial bodies through the perspective of turbulence, I had to invent a method to study turbulence itself. Part of why this was easier for me was that I was not previously trained as an expert in the field, and I did not come with any hypothesis in my mind, and I was just eager to use my scientific background and curiosity to crack a code by just following the data wherever it would lead me.

After 12 years of deep study of the data collected on the celestial bodies, I can confidently report, and I will show you shortly, that I really discovered the formula of the formation of the universe. Because I could not have cracked the origin of the universe if I did not study it through the perspective of turbulence, I ended up calling that formula the "Universe Turbulent Origin Formula."

This formula is acknowledged as the milestone that will revolutionize science forever. That same formula also helped me decode the origin of life and of chemicals. I broke it down into many books targeting different fields of expertise. That formula contains just 4 simple variables that even elementary students can understand. In fact, my 7 years old child was able to properly understand it.

I know you are eager to know the detail of this formula, but before I expose it, let me inform you that, it was after I discovered this formula that I used it to scientifically test the existence of God. As a matter of fact, I wrote this book on the existence of God in 2025 and when I started working on the origin of the universe 12 years earlier, I did not play any goal to scientifically test whether God exists or not. In other words, my journey to explain the origin of the universe and to scientifically test the existence of God did not start with a scientific hypothesis, but with a mere curiosity to check what kinds of scientific data were collected on the universe. During that search, I then discovered many variables, but the explanation of the trends that crossed my mind did not match what the scientific community was saying. Therefore, I decided to scientifically analyze those variables myself to see which secrets they may be holding. That was how I embarked into this research 12 years ago, not knowing that it would lead me to where I am today: amazing books and programs on the origin of the universe, of life, and of chemicals. You can learn more about how my programs and services can help you at Science180.com/services and Science180Academy.com.

I was shocked that the scientific formula I discovered for the first time in history landed me on a story told thousands of years ago, long before science was even born. But which story is that? Did the "Universe Turbulent Origin Formula" confirm that story? If you are ready, let's go figure out what I believe will shock you and transform your life forever. If you don't believe, keep reading, and you will see.

Before I talk to you about that formula, I will first introduce you to the details of the variables I studied and how they led me to it. Their understanding is critical for anyone who really wants to comprehend the deep scientific demonstrations I did with them. Moreover, for the sake of time and originality, I will be brief. Can the next chapter help you get closer to the answer to many questions that I know you are already asking? Before I show you what I found, let me first display something you don't want to miss about how I got here. If you want to discover the variables involved in the discovery of the aforementioned formula, so you can be better prepared to comprehend the formula itself, turn the page.

4. HISTORIC STORY OF THE REMARKABLE "UNIVERSE TURBULENT ORIGIN FORMULA" THAT EVERYBODY IS TALKING ABOUT

4.1. Can 4 simple variables and 3 celestial bodies close to us be essential to quickly and easily understand the origin of the universe and scientifically test God's existence?

If you are interested in the key raw data I used to discover the formula that enabled me to test the existence of God scientifically, you are in the right place. Indeed, everything about my scientific test of God's existence started with my work on the origin of the universe.

During my research on the origin of the universe, I came across more than 100 variables. In the process of analyzing them with my perspective of turbulence, I ended up inventing and computing more than 500 variables, which I cannot even list here. Participants of Science180 Academy (www.Science180Academy.com) have the privilege to discuss in more detail some of these variables.

The first draft I wrote with the most crucial variables was over 2,000 pages. I had to cut a lot of details before I could squeeze some of them into my book *"Turbulent Origin of the Universe"*. That book is what I call the scientific version of my book, tailored for scientists and anyone interested in the detailed scientific demonstration of the universe's formation. In that book, I used the "mother of all turbulences" to scientifically demonstrate the formation of the universe, so that scientists can understand and reorient the course of their research, teaching, and publishing, and accept the truth to live better today and forever. You can get *"Turbulent Origin of the Universe"* today to begin an incredible journey of accurately decoding the universe and changing your life forever! Learn more at Science180.com/scientific.

For the sake of time and space, I will briefly tell you what you need to know to understand the origin of the universe quickly.

Of these variables, 4 stood out the most, and without them, I could not have decoded the code of the universe. To save you time, I will focus on these 4 variables in this book. But before I explain what these variables are, let me give you their names quickly:

- Semi-major axis (which is the average distance separating a celestial body from the primary body it orbits)
- Radius (which is the straight line from the center to the circumference of the bodies)
- Orbital speed (which is the speed with which a celestial body orbits its primary body)
- Escape velocity (which indicates the minimum speed required for a body to escape the gravity of another body)

Without making you wait too long to figure it out or providing a deep understanding of the significance of these 4 variables, it would be impossible for anyone to understand the origin of the universe properly. If you don't understand what these variables mean yet, don't worry; I'll explain them to you one by one very soon. When I start doing some math later in this book, the numbers I will use for these 4 variables are those reported by NASA (National Aeronautics and Space Administration)[7].

How did I come to know that these variables are key? Indeed, as I was studying the celestial bodies in the universe, the mathematics I was doing on many variables led me to discover that, at one point during the formation of the universe, a matter having the form of fluids (behaving like a liquid like water or a gas like air) flew at a certain speed until reaching a certain place in space, where it birthed different kinds of bodies. I discovered that across the early universe, this fluid matter was organized into layers. These fluid layers were precisely moved into specific directions according to specific rules. During their movement, these fluid stacks were separated from one another based on their positions. The original fluid layers that were split into other layers are what I called "mother fluid layers", and I termed "daughter fluid layers" the bodies into which they were split. In other words, mother fluid layers were split into daughter fluid layers according to specific laws that I detailed in my book *"Turbulent Origin of the Universe"*. As these fluid layers moved, structures formed within them. Some of these structures moved in "circles," like what scientists call vortices: fluid layers wrapped around a "circle," like spaghetti wrapped around a fork. Just as inside these fluid layers structures swirled and moved from almost linear to almost circular form, it came to a point, after a certain time had passed, that the fluid layers themselves were wrapped around to form the celestial bodies, some of which are spherical.

In my book called *"How Baby Universe Was Born"*, I plainly explained these processes in a language that even children ages 7-12 can easily understand. In fact, adults are not the only ones asking the hard questions about the origin of the universe and the proof of God's existence: How was the universe formed? Did God really form it as some people believe, or did it emerge from long processes? How can we scientifically prove and break down this difficult mystery in a language that children will fully understand and like? Get the answers as you read my book, *"How Baby Universe Was Born"*, which I called the children's version of my book on the origin of the universe and life. Accurately explaining the complex formation of the universe and of life to children can be very hard in our modern world. Still, by getting *"How Baby Universe was Born"*, you will know the proven formula to help children easily understand their huge universe-origin and life-origin questions with confidence, humor, and joy. They will surely belly laugh and thank you for it! It is time to buy this pragmatic book and offer it to the children in your life today. Learn more about that book at Science180.com/children

In my book *"Turbulent Origin of the Universe,"* I spent hundreds of pages detailing the process of the formation of the universe in a scientific language that even laypeople can understand. Hence, I don't want to repeat those demonstrations here.

I invented the term "split-gathering" to describe the processes by which fluid layers in the early universe were split apart and then gathered into bodies we see today. In general, this process of split-gathering can be summarized as follows: a mother body splits into daughter bodies, which then travel a certain distance before being wrapped around to form adult bodies. Sometimes this process can repeat itself in a cascade of split-gathering before the daughter bodies can finally form. I observed these processes not only in celestial bodies but also in chemical particles and even in living things. Although they share some similarities, these processes are expressed with some nuances according to the precursors involved, whether it is a living or a non-living thing. For the sake of relevance to the objective of this book, in the rest of this chapter, I will focus on how the split-gathering happened for celestial bodies.

To be a little more scientific, I used the term "precursor" to qualify the bodies that were split to give birth to another one. In other words, the precursor of a celestial body is the body that was split and gathered to give birth to that body. In *"Turbulent Origin of the Universe,"* I showed that the precursors of bodies were made of particles that were not yet defined, and these particles underwent changes before becoming what people today call "Adult particles." In my book called *"Turbulent Origin of Chemical Particles"* known as the chemical version of my book on the universe-origin and targeting chemists, biochemists, and anyone interested in chemistry—I scientifically explained how the formation of the chemical particles and of the celestial bodies that contain them took place at the same time. In other words, chemical particles were not formed before or after the formation of the celestial bodies that contain them, or between which they are found. But as astronomical fluid layers were being split-gathered into celestial bodies, the matter in them, which at one point were fluid or fluid-like, were formed and organized into simple or complex particles according to their size, position, and movement. By fluid-like matter, I mean a matter that was not quite fluid, but that looked like a fluid. For instance, what physicists call plasma, the matter that stars are made of, is not a fluid like liquid or gas, but looks like them. With the book *"Turbulent Origin of Chemical Particles"*, the accurate deciphering and understanding of the formation of chemicals has never been profitable and easy. Hence, that great book is known as THE ultimate how-to guide for great people who want to correctly decode the origins of chemicals and positively transform their lives. Get this celebrated book today. Learn more at Science180.com/chemical.

In the aforementioned books, I showed that the way the fluid layers are split-gathered is like a gigantic seed that germinates, grows to a point before birthing branches bearing leaves, and then fruits can appear before the growth stops. In other words, I showed how the fluid layers of the precursors of celestial bodies moved and split into different fluid stacks until the last fluid stack was released. I showed that the fluid layers on top separated first, while the bottom layers are the last to split from the rest. Just as the branches on plants are not positioned at the same point, so also the daughter bodies of a mother precursor were not born at the same time, but at different times according to the duration of the processes that birthed them. Scientific details of these processes can be found in *"Turbulent Origin of the Universe."*

Before I continue, please allow me to define a few terms I will use in the next segments. Indeed, astronomical observations have shown that the universe is organized into clusters of galaxies. A galaxy is a term for a cluster of stars that are "bound" together. For instance, the closest star to the Earth is the Sun. Most stars are orbited by other celestial bodies, generally called planets and asteroids. Some planets and asteroids are orbited by other bodies called satellites. For instance, the Earth orbits the Moon. A planetary system is the system of bodies formed by a planet and its satellites. Just as satellites orbit planets and asteroids, so also some planetary systems and asteroid systems orbit a star. The Solar System is the system of bodies formed by the Sun and all the bodies orbiting it. The Sun itself belongs to the Milky Way galaxy. In my book *"Turbulent Origin of the Universe,"* I detailed how the galaxies and the celestial bodies they contain (e.g., stars, planets, asteroids, satellites) were formed. I also devoted another book to the formation of their chemical particles: *"Turbulent Origin of Chemical Particles."*

> Just as the branches on plants are not positioned at the same point, so also the daughter bodies of a mother precursors were not born at the same time, but at different time according to the duration of the processes that birthed them.

All the stars you can see with your naked eye in the night sky are just a few of the stars in the Milky Way. The stars in the sky are located at different distances from one another. Likewise, when we narrow our view to the bodies in the Solar System, the planets, asteroids, and their satellites are located at different distances from the Sun. For instance, NASA has determined that the distance separating the Sun and Mercury, known as Mercury's semi-major axis, is 57,910,000 km. Furthermore, NASA and other space agencies have confirmed that the distance separating the Sun and the Earth, known as the semi-major axis of the Earth, is 149,600,000 km. NASA and many other space agencies have confirmed that the distance between the Moon and the Earth is 384,400 km.

Did you ever ask yourself why the celestial bodies are located a specific distance from one another? Why is the Earth located 149,600,000 km from the Sun, and why is the Moon located 384,400 km from the Earth? Is it by chance? You will be shocked when I share with you the answer I got when I asked about the system of equations that I was dealing with to talk about. Before I share that with you, I will also introduce another essential variable: the orbital speed.

Celestial bodies move at different speeds, and they have different sizes. Despite being the largest body in the Solar System, the Sun does not have the highest speed in the Solar System. The fastest planet is Mercury, yet it is not the largest planet in the Solar System. Jupiter, whose radius is more than 11 times that of the Earth, moves much more slowly than the Earth. In my book *"Turbulent Origin of the* Universe," I showed that the orbital speed of the planets decreases as their distance from the Sun increases. In other words, a negative correlation exists between the orbital speed of the planets and their distance from the Sun. The same trend is also true for asteroids. Moreover, I also found that the orbital speed of the satellites and the distance separating them from their primary planet are negatively correlated. NASA proved the following orbital speeds of celestial bodies:

- Sun: 19.4 km/s
- Mercury: 47.36 km/s
- Earth: 29.78 km/s
- Moon: 1.02 km/s
- Why are celestial bodies moving at a specific speed?
- Why are they even moving at all?
- Why are the planets closest to the Sun orbiting the Sun faster than those located further away, and why are the satellites closest to their primary planet moving faster than those located further away?

Have you ever asked yourself any of the following questions?

I also found that the data collected on the radius of the celestial bodies holds a key, yet nobody had paid attention to them as we should. For instance, the Sun, which is the largest celestial body in the Solar System, has a radius more than 100 times that of the Earth. In the Solar System, the planets are located at specific distances, with the largest planets neither the innermost ones nor the outermost ones, but they are located at a certain distance between. Calculated by NASA, the radius of some key celestial bodies in the Solar System is

- Sun: 696,000 km
- Earth: 6,378.14 km
- Moon: 1,738.10 km

The question that most people, including the top scientists and even top physicists, have never asked themselves or scrutinized very well is

- Why do celestial bodies have specific mass, or what defines their size?
- Why are the largest planets located at a certain distance?
- Why is the Sun so huge, yet it moves more slowly than many bodies orbiting it?

Finally, NASA and many other space agencies across the globe have shown that for anything to leave the surface of the Sun and escape the gravity of the Sun, that thing must have a minimum speed of 617.6 km/s, which is called the escape velocity of the Sun. Likewise, NASA proved that for anything to leave the surface of the Earth and escape the pull of gravity of the Earth, that thing must have a minimum speed of 11.186 km/s (that's the escape velocity of the Earth). If you are not an expert in physics, you are probably unaware of these numbers. But if you knew these numbers, did you ever ask yourself about their origin and significance?

In Table 1, I recapitulated the raw data I addressed so far about the Sun, Mercury, the Earth, and the Moon.

Table 1: Selected facts about some celestial bodies in the Solar System[7]

Name	Escape velocity (km/s)	Radius (km)	Semi major axis (km)	Orbital speed (km/s)
Sun	617.6	696,000		19.4
Mercury			57,910,000	
Earth	11.186	6,378.14	149,600,000	29.78
Moon		1,738.10	384,400	1.02

Notes: Here, the escape velocity of the Sun is the minimum speed required for a body to escape the gravity of the Sun. Likewise, the escape velocity of the Earth is the minimum speed required to escape the gravity of the Earth. The radius of the celestial bodies is the average distance separating their center and their periphery. The semi - major axis of Mercury and of Earth is their average distance from the Sun, while the semi -major axis of the Moon is its average distance from the Earth. The orbital speed of the celestial bodies is the average speed with which they are orbiting their primary bodies.

Before I say anything else, let me recapitulate the key questions I want you to start thinking about; their answers are fundamental to explaining the origin of the universe and properly establishing the scientific test of the existence of God:
1. Why is Mercury located at 57,910,000 km from the Sun?
2. Why is the distance separating the Earth and the Sun 149,600,000 km?
3. Why is the distance separating the Moon and the Earth 384,400 km?
4. Why is the orbital speed of the Sun 19.4 km/s?
5. Why is the Earth orbiting the Sun at 29.78 km/s?
6. Why is the Moon orbiting the Earth at 1.02 km/s?
7. Why is the radius of the Sun 696,000 km?
8. Why is the radius of the Earth 6,378.14 km?
9. Why is the radius of the Moon 1,738.1 km?
10. Why is the escape velocity of the Sun 617.6 km/s?
11. What is the escape velocity of the Earth 11.186 km/s?

12. Which religion has a creation story matching these scientific evidences?

Which scientific formula can we derive from the system of these questions? Although these 12 questions I asked about the size, speed, and distance separating the celestial bodies seem simple, answering them alongside many other questions led me to what is acknowledged as one of the undeniable breakthroughs in our understanding of the origin of the universe. I invented none of the numbers I mentioned in the 12 questions above. They have been well known for more than 50 years, and some of them are even older than 300 years. In fact, for more than 200 years, the Earth's radius and the distance between the Sun and the Earth have been calculated. Would you believe me if I told you that the first person who accurately calculated that distance was an African? However, if you asked even the top physicists in the world, and even if you reviewed all the books in all the libraries in the whole world, including the classified ones, I can assure you of one thing: n minds mind of those who you will call the greatest physicists and greatest scientists whose works are used to craft the cosmological theories taught in secular schools today.

By the time you finish reading this book, you will fully understand what I mean, and I am very sure that you will thank me for what I am about to share with you very soon. All the demonstrations I did in my books on the origin of the universe, of chemicals, and of life are not needed to understand the formula that accurately tests God's existence scientifically. Therefore, in this book, I will focus on the 4 variables to explain the processes involved in the formation of the 3 celestial bodies: the Earth, the Moon, and the Sun. After 12 years of investigating the origin of the universe, I understood that we cannot say that we know the origin of the universe if we cannot master without any doubt the formation of:

- our own planet (the Earth we live on),
- the closest satellite to us (the Moon that we usually see at night), and
- the closest star to us (the Sun that we see every day).

When I surveyed the religious books in the world, I also realized that some of them that discuss the origin of the universe specifically address the formation of the Earth, the Moon, and the Sun. This suggested that it can be easy to scientifically test their creation story by rationally contrasting it with the real story that the scientific data would lead to.

Interestingly, I have scientifically discovered a groundbreaking process that explains the origin of these 3 celestial bodies, which will lighten our path to evaluating whether a God can do so or be credited as the Creator. Before I get there, let me summarize the principles involved in the formation of these celestial bodies.

4.2. Shocking story of the origin of the universe with an emphasis on the formula of the formation of the Earth, the Moon, and the Sun

I have scientifically proved in my books *"Turbulent Origin of the Universe"* and *"Turbulent Origin of the Chemical Particles"* that in the beginning, a certain particle that I termed "turbulent prima materia" appeared across a gigantic portion of space—later I will scientifically tell you where that particle came from—and then was molded into various kinds of particles that are found within and between the celestial bodies and their clusters known today. I scientifically showed that, following its mysterious appearance, the turbulent prima materia (the initial matter in the universe) was destabilized and pushed into motion, leading to the birth of turbulence within its layers, which then began moving and interacting with those in their vicinity.

The turbulence in the layers of the initial matter led to its breaking down into daughter bodies of various sizes, from the astronomical to the subatomic scales, meaning scales smaller than the atomic scale. Physics also shows that, even today, things can break down from large to small sizes through a cascade of breakups. I proved that the same happened during the formation of the universe, as gigantic mother bodies, or precursors of bodies, were broken down into daughter bodies, which in their turn were broken down into other bodies, and so on and so forth until the smallest scales at which no breakup could happen. However, the cascade of breakups I discovered differs from those previously addressed by scientists. In my book *"Turbulent Origin of the* Universe," I devoted hundreds of pages to the processes involved in this breakup of the initial matter, and if you want to know the scientific details, please refer to that book, or else we will be spending hundreds of pages here talking about things that will not interest laypeople who are also interested in this book but who lack the scientific background to understand some complex formula that even the experts are struggling with.

I discovered that during the split-gathering of the precursor of the universe, the precursor of the Solar System was formed; then, according to a specific law, it was split-gathered into the precursors caped the precursor of the Sun at a certain speed that I proved to be about what is known today as the escape velocity of the Sun. During my research, I discovered that, using the more than 300-year-old formula of the British mathematician and physicist Isaac Newton (1642–1727), physicists worldwide, including those at NASA and other space agencies, also agree that the Sun's escape velocity is 617.6 km/s. That speed is the minimum speed generally required for any body or anything to have before being able to leave the surface of the Sun. In other words, for anything to leave the surface of the Sun, it must have a speed of at least 617.6 km/s. I will get back to the significance of this number that scientists were aware of for more than 300 years, but they ignored what it really means as far as the formation of the universe is concerned.

After undergoing some changes I detailed in my book *"Turbulent Origin of the* Universe," the precursor of the Sun became the Sun. Likewise, after escaping the precursor of the Sun at about 617.6 km/s, the precursor of the bodies orbiting the Sun split and gathered into the precursors of various systems of bodies, which, upon going through some changes, became the bodies orbiting the Sun. Today, when people see the Sun and the bodies orbiting it, including the Earth-Moon system, they fail to decode the process that formed them because, among other things, they never understood the mathematics and science behind their characteristics, so they can put some equations to work to force them to tell the story of what happened in the beginning. Lucky for you, that's where my expertise and sacrifice came in: I have devoted 12 years to scrutinizing those data, and if you are interested in knowing what they said, then you are at the right place, and you are doing the right thing by devoting your attention to what I am saying.

I proved that, as the precursor of the bodies orbiting the Sun escaped the precursor of the Sun, it kept moving away, and while it traveled a distance about equal to the distance separating the Sun and Mercury, the fluid layers of the precursor of Mercury escaped. The fluid layers of Mercury's precursor underwent changes, forming the planet Mercury, which is the closest planet to the Sun. Using the distance separating the Sun and Mercury and the speed at which the precursor of the bodies orbiting the Sun moved, I was able to precisely calculate the time it took for the precursor of Mercury to escape.

After birthing the precursor of Mercury, the precursor of all the bodies orbiting the Sun continued its journey, birthing the precursors of Venus and the precursors of all the other celestial bodies (including asteroids) located between the Sun and Earth before birthing the precursor of the Earth-Moon system. Until that point, the precursor of the Earth-Moon system was a gigantic stack of fluid layers. Using the distance between Earth and the Sun, I calculated the time it took for the precursor to the Earth-Moon system to form. To get that time, I just divided the distance between the Sun and the Earth by the Sun's escape velocity, which is 617.6 km/s. Considering the data in Table 1, I showed that the precursor of the Earth-Moon system traveled about 149,600,000 km away from the precursor of the Sun before being split-gathered into the precursor of the Earth and the precursor of the Moon.

In my book "*Turbulent Origin of the* Universe," I showed that, shortly after it formed, the precursor of the Earth-Moon system split into precursors of the Earth and the Moon, each still a stack of fluid layers. I also proved that, for this split to happen, as the precursor of the Earth-Moon system was reorganizing itself after splitting from the precursor of all the bodies orbiting the Sun, the precursor of the Moon escaped the precursor of the Earth at a speed close to the Earth's escape velocity. NASA and other space agencies worldwide have confirmed that Earth's escape velocity is 11.186 km/s (see Table 1). In other words, during the formation of the Earth and Moon, after the precursor of the Earth-Moon system traveled for about 149,600,000 km (i.e., the distance separating the Sun and the Earth) away from the precursor of the Sun, the precursor of the Moon escaped the precursor of the Earth at about 11.186 km/s. I scientifically showed that after the precursor of the Moon escaped the precursor of the Earth, the precursor of the Earth was wrapped around at about the orbital speed

49

of the Earth (i.e., 29.78 km/s; see Table 1) to form the Earth, whose radius as of today is proven to be 6,378.14 km (see Table 1).

What happened to the precursor of the Moon after it escaped from the precursor of the Earth? I scientifically showed in my book *"Turbulent Origin of the Universe"* that, after splitting from the precursor of the Earth, the precursor of the Moon traveled about the distance separating the Earth and the Moon (i.e., 384,400 km, see Table 1) before reaching the position where it was wrapped around at about the orbital speed of the Moon (i.e., 1.02 km/s, see Table 1) to form the Moon, which NASA and other space agencies said is about 6,378.14 km (see Table 1). In my book *"How Baby Universe Was* Born," in a language that children ages 7-12 can understand, I showed that the wrapping around fluid layers to form spherical celestial bodies is almost like how layers of spaghetti can be wrapped around a fork.

In my book *"Turbulent Origin of the* Universe," I proved that, after the precursor of the Earth-Moon system split from the remaining precursor of the bodies orbiting the Sun, the latter continued its journey and later birthed the precursors of the other celestial bodies (planetary system and asteroid systems) located beyond the Earth in the Solar System. Because of the time it took before these precursors split from the stack of fluid layers, they were positioned at different distances. Once they were formed, they underwent processes similar to those of the Earth-Moon system's precursor before being gathered into the bodies we know today. In my aforementioned book, I also explained how the rings were formed. Because I extensively detailed those processes in those books (i.e., *"Turbulent Origin of the Universe"* and others), I will not go over them here again. Still, I will pinpoint what is required to understand the formula I used to scientifically test God's existence.

So far, I summarized the main processes involved in the formation of the Earth and the Moon. I could spend hundreds of pages detailing these stories, but for the sake of time and space, I will now explain how the Sun was formed.

I previously explained that the precursor of the Solar System was split gathered into the precursor of the Sun and the precursor of the bodies orbiting the Sun. I showed that the time that elapsed before the precursor of all the bodies orbiting the Sun escaped the precursor of the Sun should not be higher than the time it took for the fluid layers of these bodies to move away from the precursor of the Sun until they reached the position of Mercury, the innermost planet in the Solar System. Knowing the distance separating Mercury and the Sun (i.e., 57,910,000 km, which is called the semi -major axis of Mercury) and knowing the speed with which the precursor of the bodies orbiting the Sun escaped the precursor of the Sun (i.e., 617.6 km/s, m/s which is called the escape velocity of the Sun), you can easily divide that distance by that speed to determine the approximate time it took before the precursor of the Sun was fully formed with respect to the beginning of the formation of the precursor of the Solar System. After the precursor of the Sun was formed, it was wrapped at about the orbital speed of the Sun (i.e., 19.4 km/s; see Table 1) to form the Sun, whose radius, according to scientific measurements by NASA, is 696,000 km (see Table 1).

One very important thing I have not mentioned yet is the length of the fluid layers of the precursor of the celestial bodies just before they were wrapped around at the orbital speed of the celestial bodies. In other words, how long were the precursors of the Earth, of the Moon, and of the Sun just before they were wrapped at about the orbital speed of these bodies?

As I worked out the mathematics in my book *"Turbulent Origin of the Universe,"* the length of the fluid layers of the precursors of the celestial bodies just before they were wrapped around to take their spherical form was about the circumference of these bodies. In other words, the length of the fluid layers in the precursor of the Earth just before it was wrapped around to form the Earth was about the circumference of the Earth. Likewise, the length of the precursor of the Moon just before it swirled to form the Moon was about the circumference of the Moon. Finally, the length of the Sun's precursor just before it was wrapped around to form the spherical Sun was about the Sun's circumference. Those who went to middle school know that the circumference of a circle or of a sphere (something looking like an orange fruit) can be easily calculated by multiplying its radius by 2 and by pi (which is 3.14). In other words, the formula for the circumference of a sphere is

Circumference = 2 x 3.14 x Radius

Because the Earth, the Moon, and the Sun are all spherical, we can apply their radius to this formula and easily deduce their circumference. And because I also showed that the fluid layers wrapped around at the orbital speed of their daughter celestial bodies (i.e., the celestial bodies born from them), the time it took for that wrapping to occur can be calculated by dividing the circumference by the orbital speed of the celestial bodies. In other words, the time it took for the fluid layers of the Earth to be wrapped around after their split from the fluid layers of the precursor of the Moon is equal to the circumference of the Earth divided by the orbital speed of the Earth. Likewise, the time it took for the fluid layers of the Moon's precursor to wrap around just after it reached its semi-major axis (i.e., the distance between the Earth and the Moon) is equal to the Moon's circumference divided by its orbital speed. In the same manner, by dividing the circumference of the Sun by the orbital speed of the Sun, we can know the time it took for the precursor of the Sun to be wrapped around to form the Sun once it split from the precursor of the bodies orbiting the Sun. I will get back to all the calculus I did in this chapter very soon. For now, let me quickly summarize what I said in this segment in a mathematical formula.

4.3. Generic formula of the birthdate of celestial bodies

By "birthdate of celestial bodies," I mean the amount of time it took for them to be born. If you have paid attention to the story I recounted above, and if you can do a little math and science with me, you should have noticed that the formula of the duration of the formation of the Earth, the Moon, and the Sun (and of any other celestial body) involved two components:

1. the time it took before the precursor of that celestial body traveled for a certain distance away from its mother precursor or the primary body in its system before arriving at a position where it was ready to be wrapped around at a certain speed, which is the orbital speed of that celestial body as of today, and

2. the amount of time it took for the fluid layers of the precursor of that celestial body (long as about the circumference of that celestial body) to be wrapped around to form the celestial body, whose radius can be calculated using its circumference

The first part of the formula can be written as a distance (usually a semi-major axis of a certain body) divided by the escape velocity of that celestial body. For a star like the Sun, I showed that the distance to consider here is the distance separating the star from its innermost celestial body. For the Sun, I used the distance separating the Sun and Mercury, and for the speed, I used the escape velocity of the Sun. For the precursor of the Earth escaped the precursor of the Sun at about the escape velocity of the Sun. Finally for the Moon, that distance is what separates the Moon and the Earth and the escape velocity to consider is that of the Moon. For the precursor of the Moon escaped the precursor of the Earth at about the escape velocity of the Earth.

It is very important that you understand that, in the beginning, the fluid layers of the precursors of celestial bodies were not located at one single point, which can cause some people to find themselves in some form of what others call "singularity"; but the fluid layers were clustered over a large area of space that may be estimated using the amount, the distribution, and density of the turbulent materia involved, a demonstration that will inevitably lead to many misleading assumptions—scientists don't even know how big is the universe! Hence, I did not go that route, but I am content with what I did using the 4 variables I explained above.

The second part of the formula is about the time it took for the fluid layers of the precursors of the celestial bodies, long about the circumference of the celestial bodies today, to swirl around at about the orbital speed of the said celestial bodies to form that celestial body.

Therefore, the **generic formula** of the birthdate or duration of the formation of a celestial body in a Solar System with respect to a primary body is

$$\textbf{Birthdate} = \frac{\textbf{D}}{\textbf{Ve}} + \frac{\textbf{2 R }\pi}{\textbf{Vo}}$$

With

D a distance (e.g. semi major axis)
Ve an escape velocity
R the radius of the body
Vo the orbital speed of the body

This formula states that in any system of celestial bodies, the time it took for a celestial body to be formed is equal to:

- the time it took for the precursor of that celestial body to move from a certain position to another position, ready to be wrapped around and
- the time it took for the fluid layers of the precursor of that celestial body to be wrapped around at about the orbital speed of that celestial body.

Below, using the data from Table 1 that I discussed earlier in this chapter, I calculated the exact times it took for the Earth, the Moon, and the Sun to form. Applying the above formulas to the formation of the Earth, the Moon, and the Sun, I calculated what I called the birthdates of the Earth, the Moon, and the Sun. That birthdate is the total amount of time it took for these celestial bodies to be formed.

As I demonstrated in my book *"Turbulent Origin of the Universe,"* this formula applies to the formation of celestial bodies in a stellar system such as the Solar System. In that book, I explain the process of galaxy and cluster formation. In that book, I also explained why the precursors of the celestial bodies were not at a single point in the beginning. Therefore, my demonstrations exclude any notion of singularity, which could cause some people to fall into deadly technical errors. Below, I will explain what the variables in the generic formula above represent for the Earth, the Moon, and the Sun.

4.4. Birthdate of the Earth

Applying this mathematical formula to the Earth, the formula of the birthdate or the time it took for the Earth to be formed is:

$$\text{Birthdate of the Earth} = \frac{\text{Earth}-\text{Sun Distance}}{\text{Sun's Escape Velocity}} + \frac{2\pi\ \text{Earth's Radius}}{\text{Earth's Orbital Speed}}$$

Here,

- Earth-Sun Distance = Distance separating the Earth and the Sun, which is also termed the semi major axis of the Earth, and that is the distance D in the generic formula in segment 4.3 above.
- Sun's escape velocity is Ve in the generic formula
- Earth's radius is what I called R in the generic formula
- Earth's orbital speed is what is termed Vo in the generic formula
- 2π Earth's Radius = 2 multiplied by pi (i.e. 3.14) multiplied by Earth's radius = Earth's Circumference

Using the formula of the birthdate of the Earth, and the data corresponding to it in Table 1, I discovered for the first time in history that the duration of the formation of the Earth is:

Earth's birthdate = 67.29 hours + 22.42 minutes = 67.66 hours = 2.82 days

With

- 67.29 hours being the amount of time it took for the precursor of the Earth-Moon system to journey from about the precursor of the Sun to the position where the Earth is located today, and
- 22.42 minutes is the amount of time it took for the fluid layers of the precursor of the Earth to be gathered together to form the Earth

This number, 2.82 days, is what I termed Nathanael-Israel Israel's birthdate of the Earth. It is the birthdate of the Earth according to Nathanael-Israel Israel. I illustrated in Figure 1 the Earth's birthdate formula. If you want, you can verify this date by applying the data in Table 1 to the formula for the Earth's birthdate that I explained above.

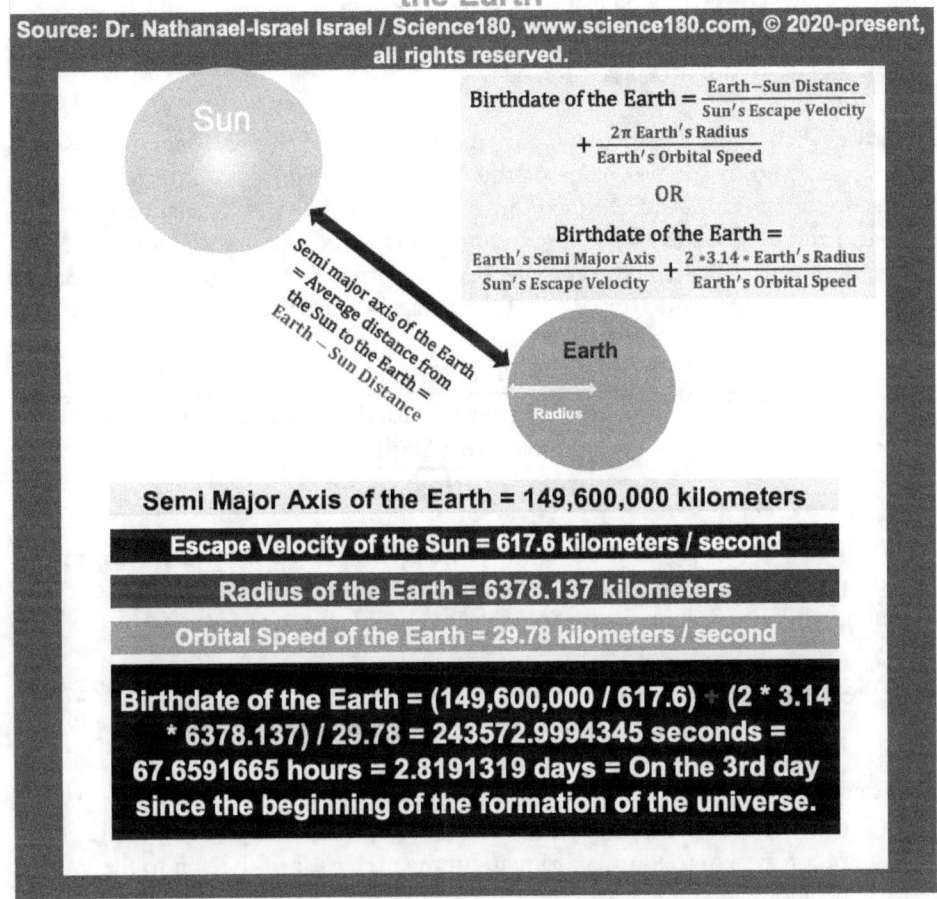

Fig. 1: Formula of the Nathanael-Israel Israel Birthdate of the Earth

$$\text{Birthdate of the Earth} = \frac{\text{Earth} - \text{Sun Distance}}{\text{Sun's Escape Velocity}} + \frac{2\pi\,\text{Earth's Radius}}{\text{Earth's Orbital Speed}}$$

OR

$$\text{Birthdate of the Earth} =$$
$$\frac{\text{Earth's Semi Major Axis}}{\text{Sun's Escape Velocity}} + \frac{2*3.14*\text{Earth's Radius}}{\text{Earth's Orbital Speed}}$$

Semi Major Axis of the Earth = 149,600,000 kilometers

Escape Velocity of the Sun = 617.6 kilometers / second

Radius of the Earth = 6378.137 kilometers

Orbital Speed of the Earth = 29.78 kilometers / second

Birthdate of the Earth = (149,600,000 / 617.6) + (2 * 3.14 * 6378.137) / 29.78 = 243572.9994345 seconds = 67.6591665 hours = 2.8191319 days = On the 3rd day since the beginning of the formation of the universe.

4.5. Birthdate of the Moon

Likewise, the formula of the time it took for the Moon to be formed after the precursor of the Moon escaped the precursor of the Earth is

$$\text{Birthdate of the Moon} = \frac{\text{Earth} - \text{Moon Distance}}{\text{Earth's Escape Velocity}} + \frac{2\pi\,\text{Moon's Radius}}{\text{Moon's Orbital Speed}}$$

Where,

- Earth-Moon Distance is the distance separating the Earth and the Moon; it is also called the semi major axis of the Moon; that is the D in the generic birthdate formula

- Earth's escape velocity is called Ve in the generic formula
- R is the radius of the Moon
- Moon's orbital speed is termed Vo in the generic formula
- 2π Moon's Radius = 2 multiplied by pi (i.e. 3.14) multiplied by Moon's radius = Moon's Circumference

However, because the precursor of the Earth-Moon system traveled a certain distance away from the precursor of the Sun before birthing the precursor of the Earth-Moon system—and the duration of that travel being the distance separating the Earth and the Sun divided by the escape velocity of the Sun—the formula of the birthdate of the Moon, or the amount of time it took for the Moon to be formed since the beginning of the formation of the Solar System, is:

$$\text{Birthdate of the Moon} = \frac{\text{Earth} - \text{Sun Distance}}{\text{Sun's Escape Velocity}} + \frac{\text{Earth} - \text{Moon Distance}}{\text{Earth's Escape Velocity}} + \frac{2\pi \text{ Moon's Radius}}{\text{Moon's Orbital Speed}}$$

If you use the data in Table 1 and the formula that I just presented concerning the birthdate of the Moon, you will see that the total amount of time elapsed since the beginning of the formation of the Solar System before the Moon was formed was

Moon's birthdate = 67.286 hours + 9.54 hours + 2.96 hours = 79.786 hours = 3.324 days

Where,

- 67.286 hours is the amount of time it took for the precursor of the Earth-Moon system to journey from about the precursor of the Sun to the position where the Earth is located today, and
- 9.54 hours is the amount of time it took for the precursor of the Moon (that escaped the precursor of the Earth at 11.186 km/s) to travel for about 384,400 km (the distance separating the Earth and the Moon) before reaching a point where it was collected into a satellite called the Moon, and
- 2.96 hours is the time it took for the precursor of the Moon to be gathered together after reaching about the position the Moon is today

This number, 3.324 days, is what I termed Nathanael-Israel Israel's birthdate of the Moon. In Fig. 2, I illustrated the formula for calculating the Moon's birthdate.

Fig. 2: Formula of the Nathanael-Israel Israel Birthdate of the Moon

$$\text{Birthdate of the Moon} = \frac{\text{Earth--Sun Distance}}{\text{Sun's Escape Velocity}} + \frac{\text{Earth--Moon Distance}}{\text{Earth's Escape Velocity}} + \frac{2\pi \text{ Moon's Radius}}{\text{Moon's Orbital Speed}}$$

OR

$$\text{Birthdate of the Moon} = \frac{\text{Earth's Semi Major Axis}}{\text{Sun's Escape Velocity}} + \frac{\text{Moon's Semi Major Axis}}{\text{Earth's Escape Velocity}} + \frac{2*3.14 * \text{Moon's Radius}}{\text{Moon's Orbital Speed}}$$

Semi major axis of the Earth = Average distance from the Sun to the Earth

Semi major axis of the Moon = Average distance from the Earth to the Moon

Semi Major Axis of the Earth = 149,600,000 kilometers

Escape Velocity of the Sun = 617.6 kilometers / second

Semi Major Axis of the Moon = 384,400 kilometers

Escape Velocity of the Earth = 11.186 kilometers / second

Radius of the Moon = 1738.1 kilometers

Orbital Speed of the Moon = 1.02316 kilometers / second

Birthdate of the Moon = (149,600,000 / 617.6) + (384,400 / 11.186) + (2 * 3.14 * 1738.1) / 1.02316 = 287260.5559841 seconds = 79.7945989 hours = 3.3247750 days = On the 4th day since the beginning of the formation of the universe.

4.6. Birthdate of the Sun

Applying the generic birthdate formula to the Sun, the birthdate of the Sun, the duration of the formation of the Sun, or the amount of time it took for the Sun to be born is

$$\text{Birthdate of the Sun} = \frac{\text{Mercury--Sun Distance}}{\text{Sun's Escape Velocity}} + \frac{2\pi \text{ Sun's Radius}}{\text{Sun's Orbital Speed}}$$

Where,

- Mercury-Sun Distance = Distance separating Mercury and the Sun, which is called the semi major axis of Mercury; that is D in the generic formula
- Sun's escape velocity is termed Ve in the generic formula
- Sun's radius is what is called R in the generic formula
- Sun's orbital speed is called Vo in the generic formula
- 2π Sun' s Radius = 2 multiplied by pi (i.e. 3.14) multiplied by Sun's radius = Circumference of the Sun

Using the data in Table 1 and the Sun's birthdate formula that I invented, I was honored to be the first person in history to discover that the time it took for the Sun to be formed was:

Sun's birthdate = 26.046 hours + 62.58 hours = 88.63 hours = 3.693 days

With

- 26.046 hours the amount of time it took before the precursor of the Sun was fully formed, which is also the amount of time it took for all the fluid layers of the precursor of the bodies orbiting the Sun to move away from the precursor of the Sun and reach the position of Mercury, and
- 62.58 hours the amount of time it took for the precursor of the Sun to wrap around to form the Sun.

This number, 3.693 days, is what I termed Nathanael-Israel Israel's birthdate of the Sun. Fig. 3 summarizes all I said in this segment about the mathematics of the duration of the formation of the Sun.

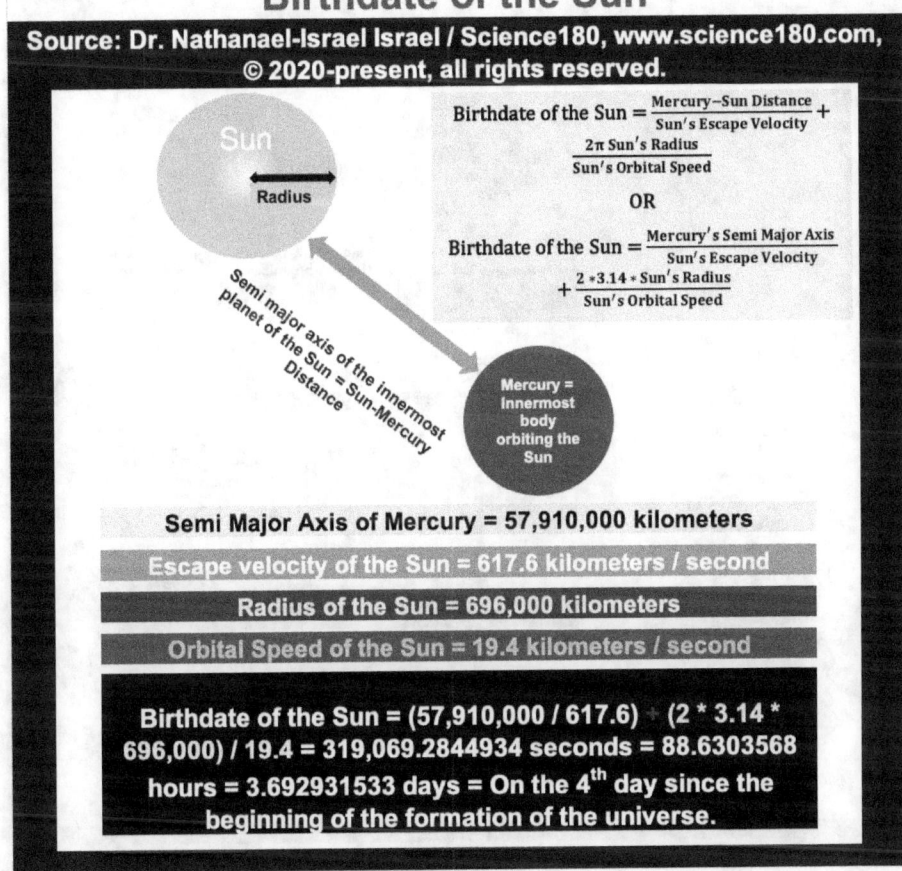

Fig. 3: Formula of the Nathanael-Israel Israel Birthdate of the Sun

Source: Dr. Nathanael-Israel Israel / Science180, www.science180.com, © 2020-present, all rights reserved.

$$\text{Birthdate of the Sun} = \frac{\frac{\text{Mercury} - \text{Sun Distance}}{\text{Sun's Escape Velocity}}}{\frac{2\pi\,\text{Sun's Radius}}{\text{Sun's Orbital Speed}}} +$$

OR

$$\text{Birthdate of the Sun} = \frac{\frac{\text{Mercury's Semi Major Axis}}{\text{Sun's Escape Velocity}}}{} + \frac{2 * 3.14 * \text{Sun's Radius}}{\text{Sun's Orbital Speed}}$$

Semi major axis of the innermost planet of the Sun = Sun-Mercury Distance

Mercury = Innermost body orbiting the Sun

Semi Major Axis of Mercury = 57,910,000 kilometers

Escape velocity of the Sun = 617.6 kilometers / second

Radius of the Sun = 696,000 kilometers

Orbital Speed of the Sun = 19.4 kilometers / second

Birthdate of the Sun = (57,910,000 / 617.6) + (2 * 3.14 * 696,000) / 19.4 = 319,069.2844934 seconds = 88.6303568 hours = 3.692931533 days = On the 4th day since the beginning of the formation of the universe.

4.7. Summary of the birthdate of the Earth, the Moon, and the Sun

Below I summarized the processes involved in how the fluid layers of the precursor of the Sun and of the precursor of the bodies orbiting the Sun split from one another and then were gathered together (Fig. 4 and Fig. 5). For clarity purposes, and so that the graph is not overloaded, I focused on the precursor of the Sun and that of a few planets.

Fig. 4: Split-gathering of the fluid layers of the precursor of the Solar System into many precursors including that of the Sun, Mercury, the Earth and the Moon

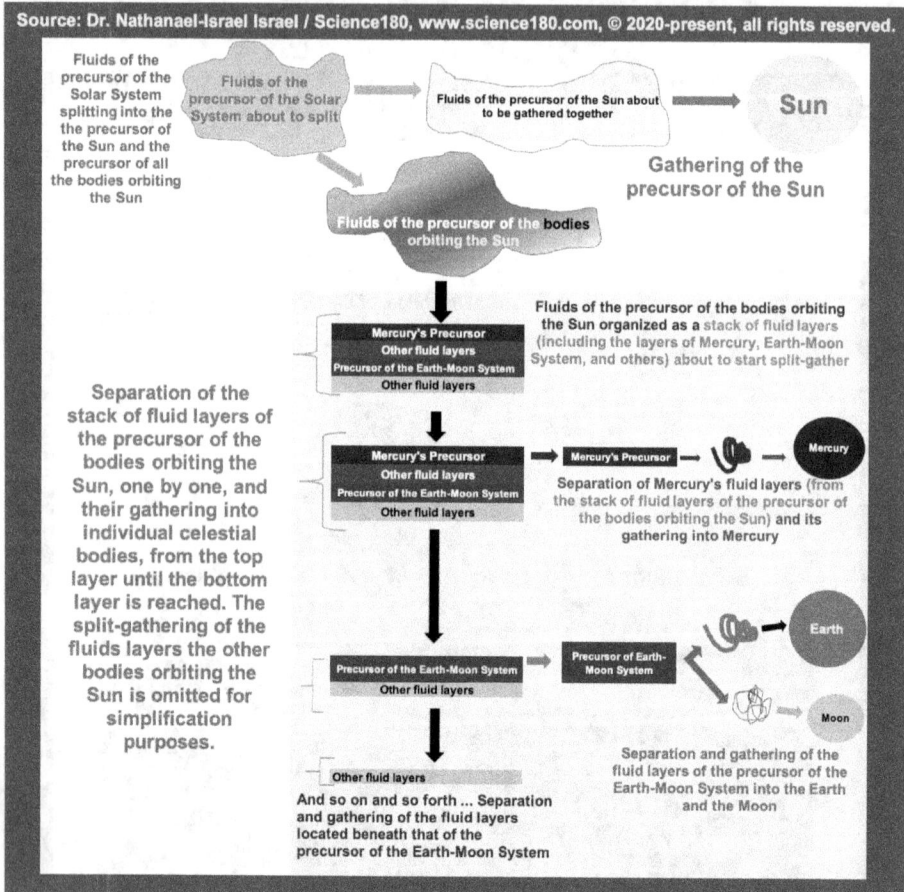

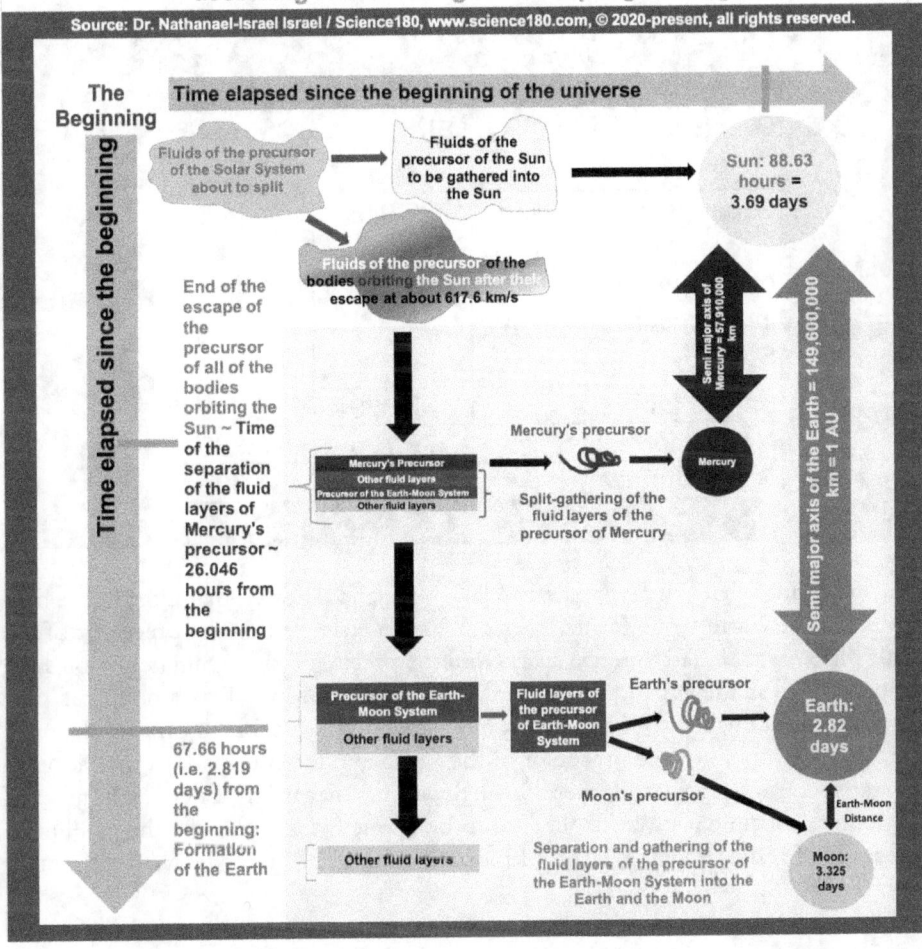

Fig. 5: Illustration of the Birthdate of the Earth, the Moon, and the Sun according to the Timing of their Split-gathering

In Table 2, I summarized the birthdate of the celestial bodies I focused on.

Table 2: Birthdate of selected major events in the history of the formation of the Earth, the Moon, and the Sun

Events	Birthdate (Hours)	Birthdate (Days)
Separation of the fluid layers of the precursor of Mercury (the innermost planet in the Solar System)	26.046	1.085
Earth's formation	67.659	2.819
Moon's formation	79.795	3.325
Sun's formation	88.630	3.693

Considering what I demonstrated so far, the science showed that most of the time it took for the Earth to be formed was due to the distance that the precursor of the Earth-Moon system had to travel away from the precursor of the Sun before birthing the precursor of the Earth and the precursor of the Moon. This amount of time accounted for a huge part of the duration of time it took for the Moon to be formed. For the time it took for the precursor of the Earth and the precursor of the Moon to be wrapped around once they were formed was not that much. In contrast, most of the duration of time it took for the Sun to be formed was caused by the amount of time it took for the large fluid layers of the precursor of the Sun to be wrapped around to form the Sun.

Does this mathematics agree or disagree with any existing theory or belief of the formation of the universe? Although this story seems weird or very strange, does it match any story foretold before the birth of science? Let's see if this story was told anywhere else that science has struggled to understand. As I said before, no such story has been published in any scientific journal. To my "surprise," before my discovery, no scientist had figured out this story that may sound bizarre to you. Therefore, the only place we can look next is the religious literature and historical books. Because secular scientists oppose religious people when it comes to the origin of the universe, it will not be surprising for us to check what the religious books say about the formation of the universe. Therefore, without wasting any more time, let's quickly review the main viewpoints of the formation of the universe of the world's main religions with an emphasis on the Earth, the Moon, and the Sun.

5. CAN ANY CREATION MYTH OR RELIGIOUS STORY IN THE WORLD SCIENTIFICALLY TEACH US ANYTHING ABOUT GOD? IF NOT, WHY?

Does the unique way I analyzed the scientific data match the creation story of any religion in the world? We cannot answer that question without first reviewing the creation narratives in the world. That is exactly what I am going to do in this chapter.

5.1. Generalities

Long before the scientific era, countless human beings lived and asked some of the same questions we ask today about our origins and the world we live in. Tentative answers were given by them, but some of the stories told to and/or by them were labeled by some people as myths, religion, heresy, unscientific, irrationality, nonsense, philosophy, confusion, ignorant fanaticism, fallacy, naivety, etc.

To account for all types of knowledge concerning God and the origin of the universe as perceived by human beings throughout the ages, and to compare what was said long before the scientific era to what scientific evidence points to today, I felt like I could not continue talking about testing God's existence without referring a little bit to the creation narratives found in the literature. Instead of detailing or ignoring the creation myths, I felt I would better summarize those found in the world's major religions.

While I will be going over these stories, just like Socrates said, "We *should follow the argument wherever it leads*," meaning letting the scientific evidence lead us to wherever they are going, including pointing to a specific religion or God if applicable. For instance, up until this point, there is no way I have said that God created the universe, but if the evidence points to that, I will not shy from using that language either.

Beforehand, some of the hot questions still unanswered, usually found when the universe's origin is discussed, and that I will also handle include:

- Is the universe a product of billions of years of evolution as some theories claimed?
- Who or what is responsible for the formation of the universe and the

"complex" order found in it?

- Because many religions claim that their God or gods created the universe, how can we (using the scientific evidence) know for sure which God or gods created the universe?
- What other religions besides Judaism and Christianity say about the origin of the universe?
- Among all the religions and philosophies in the world, which creation story matches the scientific evidence?
- Are the scientific data and the creation story in the Bible diametrically opposite?
- Was the world really created in 6 days, as the Bible claims?
- If the Biblical creation story is true, was each day of creation 24 hours or millions or billions of years?
- Is it possible to present the scientific evidence for creation without mentioning at all the Bible or any other religious book?
- Are the creation stories scientifically testable, or is creation based on supernatural processes that cannot be demonstrated scientifically?
- Can physics prove that God exists?

Before I tackle the aforementioned questions, I would like to underline that, despite their diversity, the creation stories across the world can be divided into 2 categories:

- Creation narrative of non-Judeo-Christian religions
- Creation narrative of Judeo-Christian religions

When creationism (the belief according to which the world was supernaturally created by a supreme being, God) is mentioned, people usually tend to think about the creation story particularly recounted by Judeo-Christian religions. After presenting the creation narratives, I will summarize the scientific evidence and then review if any creation narrative matches it. At that moment, I will explain if God created the universe or not, and if so, which God did it.

My goal is not to be critical or judgmental of any religion but to help you explore what is available so you can study and evaluate for yourself, objectively and dispassionately, all of the scientific evidence and arguments and, if you want, decide for yourself which religion the scientific evidence points to and which religion you can follow with conviction (if you want to) without eventually regretting it.

5.2. Creation stories

Some of the creation stories or myths have been classified into 5 categories[8]:

- Creation by the dismemberment or mutilation of a primordial being or deity, leading to the emergence of life from the corpse or dismembered parts of that originator being.

- Creation by the splitting or organization of a primordial unity (e.g., cracking of a cosmic egg or bringing forth of order from chaos). But where did the cosmic egg come from? Nobody ever answered that!
- Creation ex nihilo, which usually involved a divine being creating everything out of nothing.
- Earth diver creation, according to which a diving animal plunged to the sea floor to bring up sand or mud, which developed into a terrestrial world. But who made the diverse animal and the sea? No real answer has ever been given to that question!
- Emergence myths, according to which ancestors (or precursors) passed through a series of worlds and metamorphoses, which led to the current world.

According to some of those myths, an "embryonic" world stayed in the womb of a mother world until a new world emerged, just as a mother gives birth to a child or a seed germinates into a plant. Other myths claimed that the original world was like a union of two parents (originally bound to each other) that had to be separated or pulled apart so that offspring could be born from a sexual union between deities. Some of those myths claimed that the organs of a primordial being were detached, sacrificed, and/or transformed into celestial bodies or celestial entities (e.g., sky, earth, animals, and plants).

I hope you did not laugh at these myths, but if you do, that is ok. Some of them are really unbelievable, and you may wonder how human beings can have a brain and believe in such things! But people do believe in crazy stuff—don't they? —and they may even perceive you, who think that science is rational, as irrational. Who knows if you are not one of those who think that science is the enemy of the people!

While the above stories may appear weird or archaic to some people, carefully viewed through the glance of the level of knowledge and vocabulary available to ancient people, some of them contain an encrypted message according to which, under the influence of a Supreme Being, or something they ignore, the universe was formed through processes leading precursors of bodies to mature into more complex things as they were split and gathered together so other bodies could be born.

For the sake of time, instead of going over all the creation myths in the world, I will briefly focus this chapter on the creation stories of 5 non-Judeo-Christian religions or worldviews and the creation narrative of the Jews and Christians

- Animism
- Buddhism
- Confucianism
- Hinduism
- Islam
- Judeo-Christian religions

Before I go any deeper, I would like to mention that many people believe in none of the religious views I mentioned above, but they believe in atheism and in other forms of ideology that "deny the notion of God" or gods (idols). I already reviewed some of these viewpoints in chapter 2.

5.2.1. Animism

Animism is a belief system according to which everything in nature possesses a spiritual essence. Animists believe in no separation between the spiritual world and the physical world. They also believe that humans beings are not the only ones who have souls or spirits, but that other things and beings such as animals, celestial bodies (i.e. Earth, Moon, Sun), mountains, oceans, plants, rivers, rocks, shadows, thunder, wind, etc. also have a spirit and a soul. Hence, animists venerate the above-mentioned things. Did you ever hear about voodoo, a dominant religion in Africa, in the Caribbean, and in many other parts of the world? It is a form of animism!

Most animists believe in a creator god or a supreme being, but the name and characteristics of that god vary across cultures. Hence, animistic rituals vary across cultures, tribes, and ethnicities. Likewise, the processes by which that god or idol created the universe vary across cultures and tribes.

Although animists are found on every continent, they seem to be more dominant in Africa, Central America, South America, and Asia. Before I moved to the USA, I used to think (like most people do), as I was made to believe, that witchcraft was only found in Africa. But in the US, I saw varieties of witchcraft I could have never imagined existed in Western countries that some uninformed or evil-intended people wrongly portrayed (to some naive and conditioned people across the globe) as the "heavens" on earth or the place where the best or most advanced people live and where the so-called poor must rush to before getting some assistance, which I came to understand is sometimes a partial, wicked "return" of some of the things strategically "robbed" from the same "needy" people begging for help. Yet some witches in Western countries are clothed in scientific and political attire, while others are hidden in secret organizations, whose true visions are known only to the few initiated into these groups. The same would be found true for some American and European witches. If you tell some Africans that shrines and lodges exist in Western countries, they won't believe you!

In the end, I realized that Western witches concealed behind so-called advanced civilizations or forms of education and mentality can damage more than the African witches, some of whom I thought I left behind in Africa. I also realized that some so-called scientific theories—I said some, not all of them—are worse than some religious and fictive dogma that some people just manage to place above everything else, even above the truth, which should free them! In other words, some scientific theories about the formation of the universe are just pure myths, no better than the ancient myths I recounted in this chapter. By the time you finish reading this chapter, you will figure out some of those wrong theories and dogmas that are unfortunately promoted as the ultimate reality! To make a long story short, there is real witchcraft in all

Western nations, including the USA, where I passed the last 20+ years after having spent a few decades in Africa before moving to the country of Uncle Sam, which some people naively think or believe is (or promote to the uninformed as) the greatest paradise on earth! Western witches and animists can destroy as badly as some fetishists and totemists, who some uninformed or evil-intended people want to make the world believe are found only in Africa! I hope you are still with me. I don't want to lose you yet, because some very good stuff is coming up.

5.2.2. Buddhism

In contrast to creationist religions, there is no creator in Buddhism, a dominant Asian religion or philosophy founded in northeastern India in the 5th century BC by Siddartha Gautama (c. 563–c. 460 BC), to whom the title Buddha is given. In other words, Buddha is a title given to the founder of Buddhism, who was born in what is now Nepal.

Because Buddhism is critical of all theories on the origin and scope of the universe, debates and theories about them are usually discouraged in Buddhism to the point that belief in a creator god is often rejected[9].

5.2.3. Confucianism

Named after the Chinese philosopher Confucius (551–479 BC), its founder, Confucianism is the most dominant philosophy, political ideology, religion, social ethic, tradition, way of life, and worldview in China, where it originated. Confucianism advocates that *"the universe creates itself out of a primary chaos of material energy."* Details about the processes involved in such a creation are neither given nor emphasized. Adepts of Confucianism believe that the world is undergoing continuous ordering or configuration.

5.2.4. Hinduism

Hinduism is the dominant religious and cultural tradition of South Asia, largely in Bangladesh, India, Nepal, and Sri Lanka. Until recent years, Nepal was considered a Hindu kingdom, and Hinduism was the state religion. Because it is a collection of different intellectual or philosophical viewpoints, Hinduism does not have an inflexible common set of beliefs[10]. Consequently, Hindu texts do not provide a single undisputed and straightforward account of creation but various theories of creation[11], which leave the door widely open for evolutionism. The Hindu cosmological view of creation is grounded on a notion of repeated cycles of creation and destruction.

The Rig Veda, the oldest scripture and the backbone of Hindu philosophy, does not restrict its view on the question of God and the creation of the universe. For instance, the Nasadiya Sukta (Creation Hymn) of the Rig Veda says the following about creation[12]:

"Who really knows?
Who will here proclaim it?
Whence was it produced? Whence is this creation?
The gods came afterwards, with the creation of this universe.
Who then knows whence it has arisen?"

Talking about the creation of the primordial being, another author[13] rendered the above Creation Hymn of the Hindu philosophy as:

"Who really knows, and who can swear,
How creation came, when or where!
Even gods came after creation's day,
Who really knows, who can truly say
When and how did creation start?
Did He do it? Or did He not?
Only He, up there, knows, maybe;
Or perhaps, not even He." (Rig Veda 10.129.1-7).

"Hindu creationists believe that humans appeared or were fully formed as long as perhaps trillions of years ago[14]." This means that Hinduism supports the belief that the world is trillions of years old; hence, Hindus do not believe in the Christian version of creationism (which, as you will see shortly, advocates creation as a matter of a few days), and they oppose it. Hindus who believe in a form of creationism believe in something like a version of Christian creationism called Old Earth creationism, which some believers say erroneously supports a creation story of millions of years. Hence, Hindu creationists support the notion that the universe is older than billions of years.

5.2.5. Islam

Islam, the religion of the Muslims, was founded in the Arabian Peninsula in the 7th century AD by the Arab prophet Muhammad, also spelled Mohammed (c. 570–632). Islam is now believed by more than a billion people across the globe. The Quran (also spelled Koran), the Islamic sacred book, is believed by Muslims to be the word of God (they call him Allah) as dictated to Muhammad. Muslims believe that Muhammad is the last of the prophets meant to perfect the achievements of Abraham, Moses, and Jesus. Unlike most Christians and some Jews, Muslims deny that Jesus is God.

Just as Christians and Jews, Muslims believe in creationism, but their creation narrative significantly differs from that found in the Bible. Indeed, the Koran teaches that Allah created the heavens and the earth in six days (S. 7:54, 10:3, 11:7, and 25:59). Although the Bible also talks about 6 days of creation, some Muslims believe that a day of creation might not be an actual 24-hour solar day but an unspecified period. Therefore, many modern Muslims claim that the Koran agrees that the Earth is

68

billions of years old. As I mentioned earlier, some Christians (particularly Old Earth creationists) also share that view of a long period of creation.

The most striking differences between the Islamic narrative of creation and the Biblical account of creation can be found in the details of what the Koran and the Bible said occurred on each of the 6 days of creation.

According to the Koran, the Earth was created on Sunday (1st day of creation), and on Monday (the 2nd day of creation), plants were created on Monday (meaning Day 2), light on Wednesday (the 4th day), and heaven on Thursday (the 5th day). The Koran teaches that the stars, the Sun, the Moon, and the angels were created on Friday (the 6th day of creation).

Although it may sound as if the creation story in the Koran is the same as that in the Bible, it is important to understand that they are not. In other words, significant differences exist in the details of the creation narrative recounted in the Bible's Book of Genesis and that in the Koran. For instance, the formation of the Moon and the Sun on the 6th day is one of the many things in Islamic creationism that disagree with the Bible. A well-known source[15] revealed that "*Islamic creationism usually views the Book of Genesis as a corrupted version of God's message.*" Islam for Today[16] pointed out that the "*Koran does not contain a complete chronology of creation, and Muslim scholars do not believe in Young Earth creationism,*" a doctrine dear to many Christians[17], who believe that God created the universe in 6 literal days.

The belief in a form of creationism found in Islam may be due to the fact that the Arabs (the original adepts or fathers of Islam) and the Jews descended from Abraham; hence, Islam and Judaism are called Abrahamic religions. Most Arabs descended from Abraham's son Ishmael, whom he bore with Hagar (who was the Egyptian slave girl of his wife Sarah, whom Sarah gave to him since Sarah was barren at that time); and several years later, Abraham and his elderly wife Sarah bore Isaac, whose descendants are the Jews, let's say the Israelites.

Among believers in the Abrahamic religions, some do not accept the literal meaning of the creation story, and some accept views that oppose even what their religion teaches. Sometimes, I wonder if this is a family dispute that has gone unresolved, in which one person says, "This is what Dad told me," and the other one says, "No, this is what Dad said," or it can be that the Arabs don't want to have the same story of creation as the Jews (their number one enemy) and vice versa. Some people even think that, because the last book in the Bible was written more than 500 years before the Koran, some Biblical stories could have been copied and "twisted" by the writer(s) of the Koran so that the Islamic story of creation doesn't align 100% with the Biblical account of creation. Hence, some mismatches between the Biblical and the Islamic creation stories. But did Mohammed even really read the Bible? Later in this book, I will show whether the details of Islamic creationism match the scientific evidence I unearthed. I will get back to this serious matter very soon, I promise.

5.2.6. Evolutionist perception by the world's top religions
Proponents of evolutionism postulate that the world and everything in it are products

of processes that are billions of years old. Evolutionism has influenced many religions worldwide. In the secular world, evolutionism is considered a scientific theory, and as of 2025, before the publication of my books on the origin of the universe, of chemicals, and of life (see www.Science180.com/books), the Big Bang theory and evolution are the dominant theories of the formation of the universe and life taught in most public schools across the globe. Many private schools teach various forms of creationism on their campus, of course. My goal here is not to argue in favor of or against evolutionism, but to show you how it is perceived by the world's top religions besides Christianity. Because I already addressed how Christians perceived evolutionism in Chapter 2, I don't need to repeat it here.

Indeed, although evolutionism is not perceived by many as a religion but as a scientific theory, several religious people think that it is a religion. They argue that evolutionary assumptions are unscientific and rest on faith in the unknown. Although they deny the literalism or inerrancy of the Biblical account of creation, many adepts of the non-Judeo-Christian religions (i.e., Animism, Buddhism, Confucianism, Hinduism, and Islam) also believe in evolutionism. For instance, evolutionism is claimed as the basic premise not only of some adepts of Buddhism, Hinduism, and Confucianism. I already explained how some creationists (Jews, Christians, and Muslims) also believe in some forms of evolutionism.

In fact, in every religion, there is a group of people who believe in some form of evolutionism, although the reconciliation of their religious and scientific viewpoints can be confusing in the perspective of the things taught by evolutionism. If not, how can someone believe that the world was created in 6 days by God and, at the same time, believe that it was created by a process that took billions of years? Here comes the question of whether a creation day is really a 24-hour day or not. I will properly address that thorny question very soon.

Simultaneously, there is a group of very conservative Jews, Christians, and Muslims who, although they don't know how to demonstrate the processes involved in the 6-day creation, don't believe in evolution. Hence, a tiny group of Christians and Jews dearly hold onto the creation narrative in the Bible's Book of Genesis literally, while others, the Muslims, and the adepts of all other religions vehemently oppose the Biblical creationism in favor of evolutionism or other worldviews. How can we scientifically address these issues? Keep reading, and you will see the mind-blowing discovery I made.

5.3. Creation narrative according to Judeo-Christian religions

Although Judaism, Christianity, and Islam share the concept of creation together, significant differences exist between the Islamic creation story in the Koran and the Judeo-Christian narrative in the Bible. Christians and Jews share the same story, which is recounted in Genesis, the first book of the Bible. In other words, Judeo-Christian religions (i.e., Judaism and Christianity) drew their creation narrative from the Book

of Genesis in the Tanakh, the Hebrew Bible, or the Old Testament of the Christian
Bible.

Indeed, called Bereshit or B'reisheet (meaning "in the beginning") in Hebrew, the
Book of Genesis is also known as the First Book of Moses, the first book of the
Hebrew Bible (also known as the Christian Old Testament). Instead of paraphrasing
that Biblical creation narrative, I will quote the portion corresponding to the
formation of the celestial bodies below first; then I will elaborate on it.

> Genesis 1: 1 *"In the beginning God created the heaven and the earth. 2 And the earth
> was without form, and void; and darkness was upon the face of the deep. And the Spirit
> of God moved upon the face of the waters. 3 And God said, Let there be light: and there
> was light. 4 And God saw the light, that it was good: and God divided the light from the
> darkness. 5 And God called the light Day, and the darkness he called Night. And the
> evening and the morning were the first day. 6 And God said, Let there be a firmament in
> the midst of the waters, and let it divide the waters from the waters. 7 And God made the
> firmament, and divided the waters which were under the firmament from the waters which
> were above the firmament: and it was so. 8 And God called the firmament Heaven. And
> the evening and the morning were the second day. 9 And God said, Let the waters under
> the heaven be gathered together unto one place, and let the dry land appear: and it was so.
> 10 And God called the dry land Earth; and the gathering together of the waters called the
> Seas: and God saw that it was good. 11 ... 12 ... and God saw that it was good. 13
> And the evening and the morning were the third day. 14 And God said, Let there be lights
> in the firmament of the heaven to divide the day from the night; and let them be for signs,
> and for seasons, and for days, and years: 15 And let them be for lights in the firmament
> of the heaven to give light upon the earth: and it was so. 16 And God made two great
> lights; the greater light to rule the day, and the lesser light to rule the night: he made the
> stars also. 17 And God set them in the firmament of the heaven to give light upon the
> earth, 18 And to rule over the day and over the night, and to divide the light from the
> darkness: and God saw that it was good. 19 And the evening and the morning were the
> fourth day".* (King James Version).

The interpretation of the aforementioned creation narrative varies according to
the interpreter's religious background and has given rise to many creation theories
(most of which differ in the duration of the processes involved and in the
interpretation of certain keywords) even among those who believe in, or claim to
believe in, the Bible. Most of these theories include:

- Creation Science
- Gap creationism
- Intelligent design (proponents of this theory argue that it is not a
  creationist theory, but their opponents view them as creationists)
- Neo-creationism
- Old Earth creationism
- Progressive creationism

- Theistic Evolution (Evolutionary creationism)
- Young Earth creationism
- Science180 creationism (addressed in other books: *"Reconciling Science and Creation Accurately"* and *"Science180 Accurate Scientific Proof of God"*)

For the sake of time, I will not detail these concepts here, but based on the facts that I analyzed, I could not finish this book without saying a few words about which (if any) of those religious theories is correct. Beforehand, can I share with you my first impression after knowing what the believers and unbelievers are arguing about regarding the existence of God among themselves?

Indeed, after listening to what believers and unbelievers are saying about God, I realized that the greatest challenge to proving the existence of God is not about the proof itself, but about how to equally satisfy both those who believe in God and those who deny him. For the arguments evoked by the believers don't usually appeal to the doubters, just as those embraced by the latter don't stand in the presence of the former, who are "consumed" by faith!

Therefore, I understood that, to test the existence of God, I need to find an original way to scientifically evaluate what, according to the world's religions, God said about himself. Because we cannot do a real scientific test of the universe maker without using real numbers, I decided to put physics and math to work together to find real data in the Biblical account of creation and test what the Bible says God did during creation. I did the same test for the Islamic account of creation.

For instance, in the book of Genesis, the story of creation specifically pointed out that the formation of the Earth was completed by the 3rd day, while that of the Moon and Sun was completed on the 4th day. In the same story, the Bible says that in the beginning, "God created the heavens and the earth." In the same story, the Bible says that, at one point in its formation, the Earth was formless and that waters were separated from waters before it could be finally formed, meaning that there was a progression in the steps taken to form the Earth. How can we use real numbers and scientifically prove that the formation of the Earth was completed by the 3rd day of creation, and how can we scientifically prove that the Moon and the Sun were formed on the 4th day? For if God really created the universe, science must prove those numbers, otherwise, some rationalists cannot scientifically accept the proof.

The scientific evidence is clear enough to determine who has the right interpretation. Therefore, in the rest of this book, my main goal is to demonstrate which (if any) creation narrative best matches the scientific evidence and demonstrations I spearheaded. Because of the sensitivity of the issues and the partiality everyone has toward religion, faith, ideology, or beliefs, please read the entire book before concluding or giving up on me or my logic, should the need arise. Regardless of what you believe in or not, I agree that I have treated you and your opponent fairly. I want to address this issue of God's existence impartially once and for all. Do you want to compare the scientific evidence and the creation narratives of the formation of the universe to see what I shockingly found? If so, without any more

CHAPTER 5. CAN ANY CREATION MYTH OR RELIGIOUS STORY IN THE WORLD SCIENTIFICALLY TEACH US ANYTHING ABOUT GOD? IF NOT, WHY?

talk, let's see how the math and science we did earlier in Chapter 4 can help us figure out which creation story is right so you can have peace of mind and be happier.

6. DO YOU HAVE TO EMBRACE EVOLUTION OR DENY BIBLICAL CREATION TO SCIENTIFICALLY PROVE THAT GOD CREATED THE UNIVERSE IN 6 LITERAL DAYS?

- Is there any need to prove the Biblical creation to be true?
- Did Biblical creation really undermine the scientific enterprise?
- Are Biblical creationists really explaining creation as they say?
- What if the Bible scientifically teaches something about creation that smart scientists ignore and that Christians refuse to demonstrate to win over rationalists and freethinkers?
- Is it a waste of time to attempt to prove the Bible is true by means of science or historical investigations?
- Can anyone scientifically solve the most-asked questions about the Biblical narrative of the creation of the universe, revealed about 3500 years ago?
- Why is water confirmed on most celestial bodies in the solar system?
- Is there really a God who created the universe and everything in it?
- Were all celestial bodies really formed completely by the end of the 6th day of creation, as the Bible says?
- Are the creation days 24 hours each or millions of years long?

Have you ever wondered, "How can I scientifically understand the Biblical creation quickly without falling into the trap of taking sides between science and faith?" Have you ever struggled with the interpretation of some verses in the Genesis story of creation, not knowing what to do or what sense to make of them without doubting God or checking your brain at the door? What if this chapter can help you get closer to the answer to those questions and many more? Well, I want to share a groundbreaking approach with you.

As you know, it took me 12 years of investigation and writing thousands of pages before I understood the formation of the universe as I presented in my books (see www.Israel120.com/books). Here, for the sake of space and time, I will pinpoint a few points relevant to testing the creation narratives I presented earlier. Afterwards, I will scrutinize whether any narrative matches scientific data. Attention will be given to evidence related to the timing of the:

- split-gathering of fluids in the precursors of celestial bodies
- formation of the Earth
- formation of the Moon
- formation of the Sun

6.1. Recap of the formation of the Earth

Like I proved in chapter 4 and in more detail in my book *"Turbulent Origin of the Universe,"* after the precursor of the bodies orbiting the Sun escaped the precursor of the Sun at about 617.6 km/s (i.e., the escape velocity of the Sun, see Table 1), it was organized as a stack of fluid layers. In that stack of fluids, the precursor of the Earth-Moon system was embedded somewhere and had to "wait" until all the fluid layers of all the bodies located above it (e.g., the precursor of Mercury, of Venus, and of many asteroids located between the Sun and the Earth) split and were removed from above the precursor of the Earth-Moon system before the latter could (at its turn) split and gather itself into the precursor of the Earth and the precursor of the Moon. The distance traveled by the precursor of the Earth-Moon system before it split from the fluid layers below it was about the semi-major axis of the Earth (i.e., the average distance separating the Sun and the Earth), 149,600,000 km. That distance is called the astronomical unit (symbolized as 1 AU). In other words, the speed with which the precursor of the Earth-Moon system traveled from the precursor of the Sun to the point of its split from the stack of fluid was about the escape velocity of the Sun, 617.6 km/s. Dividing the aforementioned distance by that speed, the time it took for the precursor of the Earth-Moon to travel that distance is about 67.29 hours (i.e., 2.804 days).

I showed that, after the fluid layers of the precursor of the Earth-Moon system split from the stack and became "independent" of the other fluid layers above and below them, it did not take long before they split into the precursors of the Earth and the Moon. Considering the trends that I obtained for the lifespan of the precursor of all the planetary systems in the Solar System, I showed in my book *"Turbulent Origin of the Universe"* that the precursor of the Earth-Moon system split very quickly in a matter of a few minutes.

After the precursor of the Earth-Moon system split into the precursor of the Earth and the precursor of the Moon, the precursor of the Earth was wrapped at about 29.78 km/s (i.e., the orbital speed of the Earth). In other words, the fluid layers of the precursor of the Earth were moved at about 29.78 km/s (orbital speed of the Earth) and then were wrapped around or collected all around to form a spherical planet called

Earth, whose equatorial radius is about 6378.137 km. Considering the radius of the Earth, I calculated the circumference of the Earth (that is, the maximum distance a fluid parcel could have taken to move all around the Earth) as

$$\text{Circumference} = 2 * 3.14 * \text{Earth's Radius}$$
$$\text{Circumference} = 2 * 3.14 * 6378.137 \text{ km} = 40{,}054.7 \text{ km}$$

By dividing that circumference (which is a distance) by the speed with which the gathering together of the fluid layers of the precursors of the Earth occurred, I showed that the duration of the gathering of the fluid layers of the precursor of the Earth together around its circumference was about:

$$\text{Circumference / Orbital speed} = 40054.7 \text{ km} / 29.78 \text{ km/s} = 1345.02$$
$$\text{seconds} = 22.42 \text{ minutes}$$

This means that in 22.42 minutes after they reached a position about the semi-major axis of the Earth, all of the fluid layers of the precursor of the Earth were gathered together into the spherical Earth we know today.

Adding together the time it took for the fluid layers of the precursor of the Earth-Moon system to move from about the precursor of the Sun to the orbit of the Earth (67.29 hours) and the time it took for them to be gathered together around the circumference (22.42 minutes), I showed that the Earth's birthdate, meaning the duration of the formation of the Earth, is:

$$\text{Earth's birthdate} = 67.29 \text{ hours} + 22.42 \text{ minutes} = 67.66 \text{ hours} =$$
$$2.82 \text{ days}$$

This proved that the formation of the Earth was really completed on the 3rd day of creation, just as the Bible's Book of Genesis said (Genesis 1:9-13):

> Genesis 1:9 *"And God said, Let the waters under the heaven be gathered together unto one place, and let the dry land appear: and it was so. 10 And God called the dry land Earth; and the gathering together of the waters called the Seas: and God saw that it was good. 11 … 12 … and God saw that it was good. 13 And the evening and the morning were* **_the third day_***"*.

As a reminder, I illustrated this mathematic in Fig. 1 in chapter 4.

6.2. Recap of the formation of the Moon

The history of the formation of the Moon is connected to that of the Earth. For

instance, some of the events that I described for the formation of the precursor of
the Earth-Moon system are part of the story of the formation of the Moon. Indeed,
once the precursor of the Earth-Moon system split from the stack of fluid layers of
the precursor of the bodies orbiting the Sun, it went through its own split-gathering
to yield the precursor of the Earth and the precursor of the Moon. The precursor of
the Moon escaped the precursor of the Earth at about the escape velocity of the Earth
(11.186 km/s). As the fluids of the precursor of the Earth were being gathered
together into the Earth, the fluids of the precursor of the Moon were moving away at
about 11.186 km/s and traveled for about 384,400 km (i.e., the semi-major axis of the
Moon = average distance separating the Earth and the Moon) before reaching a point
where they were in a position to be collected into a satellite called the Moon. By
dividing that distance by the speed of the travel, I obtained the duration of that travel
as 9.54 hours:

$$384,400 \text{ km} / 11.186 \text{ km/s} = 34364.38 \text{ seconds} = 9.54 \text{ hours}$$

Another way to explain this number is that, after traveling 9.54 hours at a speed
of 11.186 km/s away from the precursor of the Earth, the precursor of the Moon
could collect itself into a satellite. Considering what I said above about the timeline
of Earth's formation, by the time the precursor of the Moon was gathering into the
Moon, Earth's formation was already complete.

Just as I did for all of the celestial bodies I studied, the gathering together of the
fluid layers of the precursor of the Moon occurred at the orbital speed of the Moon
(i.e., 1.02316 km/s). Just like I did for the Earth and the Sun, the time it could have
taken for the fluids of the precursor of the Moon to be gathered into the spherical
Moon (radius = 1738.7 km) was the circumference of the Moon divided by the orbital
speed of the Moon. Based on the radius of the Moon, the circumference of the Moon
is:

$$\text{Moon's circumference} = 2 * 3.14 * 1738.1 \text{ km} = 10,915.27 \text{ km}$$

Dividing distance by the orbital speed of the Moon, I showed that the time it took
for the Moon to be gathered together after reaching its semi-major axis was

$$10,915.27 \text{ km} / 1.02316 \text{ km/s} = 10,668.2 \text{ seconds} = 2.96 \text{ hours}$$

With respect to the precursor of the Earth-Moon system, the time elapsed before
the Moon was fully formed is the sum of the time it took for the precursor of the
Moon to escape the precursor of the Earth-Moon system and reach the orbit of the
Moon (9.54 hours) and the time it took for the fluid layers of the precursor of the
Moon to be wrapped around to form a spherical body having a radius like that of the
Moon (2.96 hours):

$$9.54 \text{ hours} + 2.96 \text{ hours} = 12.5 \text{ hours}$$

In other words, the Moon was formed 12.5 hours after the split of the precursor of the Earth-Moon system into the precursor of the Earth and that of the Moon. Because it took 67.286 hours for the precursor of the Earth-Moon system to move from the precursor of the Sun and reach a position in space where it split from the stack of fluids of the precursor of the bodies orbiting the Sun and then split into its daughter bodies (the precursor of the Earth and that of the Moon), the duration of the formation of the Moon with respect to the Sun must consider all the time that occurred before the precursor of the Moon was even born. To put it another way, the birthdate of the Moon, meaning the total amount of time that elapsed since the beginning of the formation of the Solar System before the Moon was formed, was

$$\text{Moon's birthdate} = 67.286 \text{ hours} + 12.5 \text{ hours} = 79.786 \text{ hours} = 3.324 \text{ days}$$

Right on point, the Moon was formed on the 4th day, just like the Bible said more than 3500 years ago (Genesis 1:14-19):

> Genesis 1:14 "*And God said, let there be lights in the firmament of the heaven to divide the day from the night; and let them be for signs, and for seasons, and for days, and years: 15 And let them be for lights in the firmament of the heaven to give light upon the earth: and it was so. 16 And God made two great lights; the greater light [the Sun] to rule the day, and the lesser light [the Moon] to rule the night: he made the stars also. 17 And God set them in the firmament of the heaven to give light upon the earth, 18 And to rule over the day and over the night, and to divide the light from the darkness: and God saw that it was good. 19 And the evening and the morning were **the fourth day**".*

In Fig. 2 (see Chapter 4 of this book), I summarized the math I just did about the birthdate of the Moon.

6.3. Recap of the formation of the Sun

As I have demonstrated in Chapter 4 of this masterpiece you are reading, after the precursor of the Solar System was born, it did not take a lot of time before it split into the precursor of the Sun and the precursor of the bodies orbiting the Sun. The precursor of the bodies orbiting the Sun escaped the precursor of the Solar System or split from the precursor of the Sun at about the escape velocity of the Sun (617.6 km/s). Before such a split was consumed so that the precursor of the Sun could be ready to start gathering its fluid into a spherical body, the precursor of the bodies orbiting the Sun traveled a distance no higher than the average distance between the

Sun and Mercury (Mercury being the closest planet to the Sun). That distance (also termed as the semi-major axis of Mercury) is equal to 57,910,000 km = 0.387 AU. Considering the aforementioned distance and speed, the maximum time it took for the fluid layers of the bodies orbiting the Sun to escape the precursor of the Sun is about:

$$57,910,000 \text{ km} / 617.6 \text{ km/s} = 93766.2 \text{ seconds} = 26.046 \text{ hours}$$

That is what I called the escape time of the precursor of the Sun.

Because the precursor of the Sun could have waited for the precursor of all the bodies orbiting it to escape its surface before it could start gathering together all its fluid layers, the time I calculated above could have elapsed before the precursor of the Sun could have started collecting its fluid together. Once the precursor of the bodies orbiting the Sun escaped, the fluid layers of the precursor of the Sun moved at about the orbital speed of the Sun, 19.4 km/s, and were gathered together by kind of spiraling, winding up, or wrapping up movement at the previously mentioned speed (19.4 km/s) to form a spherical and massive star called the Sun, whose radius is 696,000 km.

In my book *"Turbulent Origin of the Universe,"* I described why the precursor of the Sun formed a star but not a solid body like the terrestrial planets. Knowing the radius of the Sun, I estimated its circumference as

$$\text{Sun's Circumference} = 2 * 3.14 * 696,000 \text{ km} = 4,370,880 \text{ km}$$

That 4,370,880 km is about the distance a fluid parcel could have traveled to complete one turn around the Sun. Dividing the circumference of the Sun by its orbital speed allowed me to estimate the time it could have taken for the precursor of the Sun to gather together around all its fluids to form the spherical body like the Sun after the fluid of the bodies orbiting the Sun escaped the precursor of the Sun:

$$4,370,880 \text{ km} / 19.4 \text{ km/s} = 225,303 \text{ seconds} = 62.58 \text{ hours}$$

Based on what I already demonstrated above, the total amount of time it took for the Sun to form is the sum of:

- 26.046 hours (escape time of the precursor of the Sun): the time it took for the fluids of the precursor of the bodies orbiting the Sun to clear the way or to escape the precursor of the Sun and
- 62.58 hours: the time it took for the fluid layers of the precursor of the Sun to be wrapped around to form a spherical star called the Sun.

That amount of time, which I called the Nathanael-Israel Israel birthdate of the Sun, is

Sun's birthdate = 26.046 hours + 62.58 hours = 88.63 hours = 3.693 days

In other words, according to the scientific evidence that no human being has properly analyzed until my days, the days of Nathanael-Israel Israel, the Sun was formed on the 4th day since the beginning of the formation of the Solar System, which I demonstrated in my books *"Turbulent Origin of the Universe"* and *"Reconciling Science and Creation Accurately,"* and is also the beginning of the formation of the universe. This scientific evidence is perfectly aligned with the Biblical creation narrative, according to which the Sun was formed on the 4th day (Genesis 1:14-19):

> Genesis 1:14 *"And God said, let there be lights in the firmament of the heaven to divide the day from the night; and let them be for signs, and for seasons, and for days, and years: 15 And let them be for lights in the firmament of the heaven to give light upon the earth: and it was so. 16 And God made two great lights; the greater light [the Sun] to rule the day, and the lesser light [the Moon] to rule the night: he made the stars also. 17 And God set them in the firmament of the heaven to give light upon the earth, 18 And to rule over the day and over the night, and to divide the light from the darkness: and God saw that it was good. 19 And the evening and the morning were **the fourth day**"*.

In Fig. 3, I illustrated this math of the birthdate of the Sun.

6.4. Is there really a God who created the universe and everything in it? If yes, which God is that?

I understand that most of the religions or ideologies in the world try to claim that their creation story is true and that of others is wrong. In most of the previous segments of this chapter, I explained how, according to the historic scientific evidence, the Biblical narrative of creation is the only true one. For instance, the Islamic narrative is incorrect, for its dating of the creation of the celestial bodies (Sun, Moon, and Earth) does not match the scientific evidence, although it also claims the 6 days of creation. In other words, the details of the events that occurred in each of the 6 days of the Islamic creation story narrated in the Koran did not corroborate the scientific evidence as the Bible did for almost 3500 years. Furthermore, before my work, no Muslim had ever scientifically demonstrated if, how, and why their creation story of 6 days is true or not. Unlike the Bible, which presents a clean and historical account of creation, the Koran does not have a clear account of creation, but many references are made here and there about creation. Although the Koran talks about God as Allah, the scientific evidence proved that Allah did not create the universe. I will revisit this question about the Islamic account of creation later in this book with more shocking details that you don't want to miss.

CHAPTER 6. DO YOU HAVE TO EMBRACE EVOLUTION OR DENY BIBLICAL CREATION TO SCIENTIFICALLY PROVE THAT GOD CREATED THE UNIVERSE IN 6 LITERAL DAYS?

In *"Reconciling Science and Creation Accurately,"* I explained other key facts that the Bible mentioned concerning the creation of the universe, all of which are true.

In the Biblical story of creation, Moses did not say that the universe was formed by chance, but he clearly stated that God (the God of Israel) created it. Because the story that Moses told millennia ago is scientifically found to be true, as demonstrated in my writings, we must also accept the fact that Moses said that God is the Creator. Furthermore, as found in the literature, Moses has also authored 4 other books in addition to the Book of Genesis:

- Exodus
- Leviticus
- Numbers and
- Deuteronomy

Together, these 5 books of Moses are called the Torah (in Hebrew), meaning "Law" or "Teaching," also known as the Pentateuch. Unfortunately, for a long time, some people have considered these books as "primarily mythological rather than historical." Yet, some Biblical literalists have been considering them as actual history. In those books, Moses revealed on many occasions that there is only one God, the God of Israel, referred to in the Biblical creation story as Elohim, but known to the Jews (including the Black Jews or Hebrew Israelites) by many other names, including Adonai, El, Yah, Yahweh, and Jehovah, a name that the Jews have not taken lightly and which they avoid writing as such; hence, its codification in the tetragrammaton! Due to the purpose of this book, I will not go any deeper into theological details and discussions about the nature, attributes, or characteristics of God, the Creator, but I addressed them in other books I previously mentioned. Countless well-written books in the literature have devoted resources to address who the God of Israel is. Below are some Biblical references about what God said about Himself and what many people testified about Him for thousands of years:

- *"You shall have no other gods before Me"* (Exodus 20:3)
- *"I am the LORD, that is my name! I will not give My glory to anyone else, nor share My praise with carved idols"* (Isaiah 42: 8)
- *"I am He. Before Me, there was no God formed, nor shall there be after Me. I, even I, am the LORD, and besides Me there is no savior"* (Isaiah 43:10-11)
- *"I am the first and the last, and there is no God except Me"* (Isaiah 44:6)
- *"I am the LORD, and there is none else, there is no God beside Me"* (Isaiah 45:5)
- *"Unto you, it was showed, that you may know that the LORD is God, there is none else beside Him"* (Deuteronomy 4:35)
- *"Know therefore this day, and consider it in your heart, that the LORD He is God in heaven above, and upon the earth beneath: there is none else"* (Deuteronomy 4:39)
- *"Hear, O Israel, the LORD our God, the LORD is one!"* (Deuteronomy 6:4)
- *"That all the people of the earth may know that the LORD is God; and there is no other"* (1 Kings 8:60)

- *"There is none like You, O LORD, and there is no God besides You, according to all that we have heard with our ears"* (1 Chronicles 17:20)
- *"For You are great, and do wondrous things: You are God alone"* (Psalm 86:10)
- *"You, even You are LORD alone; You have made heaven, the heavens of heavens, with all their host, the earth, and all things that are therein, the seas, and all that is therein, and You preserved them all; and all the host of heaven worshipped You"* (Nehemiah 9:6)
- *"For us, there is one God, the Father, from whom are all things and for whom we exist, and one LORD, Jesus Christ, through whom are all things and through whom we exist"* (1 Corinthians 8:6)
- *"For there is one God and one mediator between God and mankind, the man Christ Jesus"* (1 Timothy 2:5-6)

In other words, although people across the globe and throughout history have labeled their idols as "gods," there is only one true God, the God of Abraham, Isaac, and Jacob, the God of Israel who created the heavens and earth and everything within and between them. Any other being called a "god" or even "God" besides Him is a false god or an idol. It is a huge mistake to think or pretend to unite people across the globe by claiming that there are many gods or that the being(s) called "god" in all the religions of the world are the same. In other words, although there are many "gods" or "idols," there is ONLY ONE GOD, the God of Abraham, Isaac, and Jacob. Although many Jews, Arabs, and Gentiles claim to serve the same God or descend from the same father, Abraham, the ONLY TRUE GOD is the one worshipped by Christians and Messianic believers in Yeshua HaMashiach (Jesus Christ, who, according to the Bible, is the way, the truth, and the life), without whom no one can ever see or know God. If you read the Gospel of John, you will see that even most of the Jews, whose ancestors God revealed himself to a lot, don't know Him. If not, why did the Pharisees and Sadducees plot to kill Jesus (Yeshua), whom the Messianic Jews are now massively accepting as the Savior? My friends, I know some of you may be frustrated at what I am saying here, but I deeply felt like I must say a few central things at this point not only to help someone find the truth but also to wash my hands of any guilt of not sharing the truth about the origin of the universe. Like the Bible says, salvation can be found only in Jesus (Yeshua HaMashiach).

To wrap this up, I would like to say that, although some people "believe" that God created the world, some wrongly accept as truth that He did so, not out of nothing or using His word only as the Bible claims, but they wrongly say that God used evolution (over a period of billions of years) before forming the world we know as of today. However, the evidence in this book proved that creation was indeed done in a matter of literal days, each consisting of only 24 hours. Furthermore, it is important to notice the difference between the creation and the formation of something. Derived from the Latin phrase "ex nihilo," which means "out of nothing," "creatio ex nihilo" means "creation out of nothing," whereas "creatio ex materia" stands for "creation out of some pre-existent" materials.

Applying these terms to the formation of the celestial bodies, I can say that to some extent, the appearance of the "turbulent prima materia" (the original matter in the universe) is a "creatio ex nihilo," whereas the formation of all the bodies by molding the "turbulent prima materia" is a "creatio ex materia." But to be correct, God used His words to create (John 1:1); therefore, it is incorrect to talk about "creation out of nothing." But because before He spoke the word; there may have been "nothing" besides Himself, who is self-sufficient; it may be ok to talk about "creation out of nothing." Even so, before He spoke, God had a thought and a plan to create, which is something. Putting this philosophical idea in a different way, it would be better to say that God created everything, not out of nothing, but out of Himself, using His word, which brought forth an initial matter, which He used to mold everything in the universe according to His plan, which irrevocably established the foundation of everything by the end of the 6th day, meaning that although certain things (such as the Earth, the Moon, the Sun, and many other celestial bodies) were fully formed by the end of the 4th day, others were still being molded and they finished their formation a few days later, even days after the 7th day of rest, so that order can be seen in the universe according to His will and design, which is visible but encrypted in the world, which He created for His own pleasure and by His own will...

> Why does God, the uncreated Creator, have no beginning and no end? What was God doing before creation?

Considering the information, I have in the books I wrote on the origin of the universe (www.Israel120.com/books), I know that some people may still have important questions needing a correct answer:

- What was there before the beginning of the created world?
- What and where was the beginning?
- Are there any other types of space beyond the limit of the space known to humankind?
- Why does God, the uncreated Creator, have no beginning and no end?
- Why is God fire?
- Why is God light?
- Where was God before creating the world and during creation?
- What was God doing before creation?
- What is time?
- Was there a "zero time" just before the birth of the precursor of the universe?
- Did any time exist before the time known as of today began?
- Is the Earth mentioned in Genesis 1:1 the completely formed Earth known as of today?

- How did time start before the completion of the formation of the Sun on the 4th day?
- Because the Sun was not created until Day 4, how could the first three days have been ordinary days?
- Are the heavens and earth mentioned in Genesis 1:1 the same as the heaven and earth today?
- Do the heavens mentioned in Genesis 1:1 include the Heaven that some people say is the dwelling place of God?
- Does God really live in the heaven mentioned in Genesis 1?
- Where was God living before creating the heaven?
- Where is the location of heaven?
- Why, according to Genesis 1, was the Spirit of God moving on the surface of the waters?
- Was the throne of God created or not?
- How were water and light formed?
- Who are angels, and why, when, and how were they formed?
- Why did God create the universe in 6 days?
- What is the shape of the universe?
- When will the world end?
- Why did God hide some secrets and prevent some human beings and even angels from discovering them quickly?

What if other books I wrote can help you accurately answer those questions with an originality and tune you will not find anywhere else? Indeed, in my books *"Reconciling Science and Creation Accurately"* and *"Origin of the Spiritual World,"* I better elaborated on many things, including these questions that I cannot fully address in this book due to space constraints and the nature of its objective. For the demonstration I introduced in this book is just a glimpse at the gigantic iceberg of the Biblical truth concerning the duration and the processes of the formation of the universe. Those who attend Science180 Academy (www.Science180Academy.com) also get all of those questions answered. And if you are interested, you can register to meet me there.

Finally, what I found interesting while writing this book is that many people refuse to believe in God, yet they delight in inventing companies and/or working for organizations that seek to send human beings to space, even to have human beings colonize other planets, and I wonder if they even know how they were formed and how costly (maybe uncomfortable and impossible) it can be for human beings to live there. As I was reflecting on my findings, I was reminded of how interesting the timing of the discovery and publication of my works is compared to recent years during which billionaires are racing to send people into space. In fact, from 2020 to 2025, when I was working on the 9 books I published in 2025, and when I was thinking about this issue, many millionaires and even billionaires visited space on board private spacecraft. Are we going to continue enjoying the universe without properly thinking about the Creator and what He expects from us?

6.5. Recap of the split-gathering of fluid layers

Considering what I wrote on the birthdate of the Sun in chapter 4, I showed that it took about 26.05 hours maximum for the precursor of the bodies orbiting the Sun to split from the precursor of the Sun and reach about the position of Mercury (which, as of 2025, is the main innermost body orbiting the Sun). Obtained by dividing the distance separating the Sun and Mercury by the escape velocity of the Sun, these 26.05 hours are what I called the escape time of the precursor of the Sun.

In the Biblical account of creation, it is mentioned that waters started dividing from waters on the second day, which means after the first 24 hours. Because no (major) celestial body is closer to the Sun than Mercury, the afore-calculated 26.05 hours means that waters were not yet separated from other waters (in the Solar System) before the beginning of the second day. This concurs with the Biblical account about the separation of fluids on the 2nd day.

The Book of Genesis talks a lot about waters referred to as "Mayim" in Hebrew. In the days when Moses was recounting the Biblical narrative of creation, a lot was not known about chemical elements yet, and much less about fluid dynamics. It took many centuries afterward before the first chemical element was discovered (scientific details about the formation of chemicals can be found in my book "Turbulent Origin of Chemical Particles").

Therefore, the term "waters" used in the creation narrative of Genesis 1 should not be limited to a pure water only. It may have been a water or a water-like fluid mixed with other chemical elements and compounds (and/or precursors of chemical elements and compounds) as they were being formed. In other words, what Moses (the author of the Biblical book of Genesis) was referring to as waters could not have been just pure water consisting of hydrogen and oxygen molecules only, but a water mixed with other chemical elements and/or their precursors. For many other chemical elements could have been formed by the time the Earth was formed. In the days that Moses revealed the creation narrative, nothing was known about chemistry. Even more than 1000 years after the writing of Genesis, the early Greek scientists still classified everything in the universe using just water, fire, air, and soil. In other words, Moses (who lived about 1500 years BC) could not have done a better job to express the fluid-like precursor of the Earth.

Because, as far as the celestial bodies in the universe are concerned, the Bible specifically named the Earth, the Sun, and the Moon, and because, according to the scientific evidence, all of those bodies were formed by Day 4 (see what I said about the birthdate of celestial bodies in chapter 4), it is fair to say that the split of waters mentioned on the 2nd day of the Biblical creation was at least about the fluids of the precursors of some bodies in the Solar System. As I detailed in the chapter on the semi-major axis timescale in my book *"Turbulent Origin of the Universe,"* the fluid layers of the precursor of Mercury, Venus, and the Earth-Moon system, and those of the asteroids between them, were actively divided on the second day. In that book, I showed that, during these first 2 days, the fluids of the precursors of celestial bodies in other stellar systems were going through similar processes of split-gathering according to their environmental conditions. Therefore, on the 2nd day, waters were really being divided or separated from waters just like the Bible said (Genesis 1:6-8):

> Genesis 1:6 *"And God said, let there be a firmament in the midst of the waters, and let it divide the waters from the waters. 7 And God made the firmament, and divided the waters which were under the firmament from the waters which were above the firmament: and it was so. 8 And God called the firmament Heaven. And the evening and the morning were the **second day**"*.

In *"Turbulent Origin of the Universe,"* I also demonstrated how the sky and the atmosphere of celestial bodies were formed during the processes that split and gathered the precursor of the celestial bodies.

According to the Bible, the waters that were the precursor of the Earth were divided or separated into different compartments (Genesis 1:6-8) and then the sky was created in the opened space. During the formation of the sky, a portion or layer of water was moved above the sky while another portion was positioned below the sky. The Earth was formed using the portion of the water that was below the sky.

Although the story in Genesis 1 did not specifically say the outcome of the waters that went above the sky, other references suggest that the water above the sky could have been the water that rained on Earth in the days of Noah when the Great Flood occurred. God had to open the window of heaven for the water to rain down for 40 days and 40 nights (Genesis 7:11). I elaborated on that in my book *"Reconciling Science and Creation Accurately."*

The Bible did not say whether or not, during the great flood (in the days of Noah), God poured out all the water that was moved above the sky onto the Earth. However, the flood story recounts that, at the end of the 40th day of rain, God shut the window of heaven and the rain stopped. This implied that during the early years after creation, the Earth could have been surrounded by layers of water, which could have been similar to how layers of the ozone and other gases surround the Earth today. It is possible that the mass of water above the sky is referring to a fluid mass even beyond the Solar System, for the location surrounding where the Sun was formed or where it exists today (called Rakia in Hebrew) seems to express the space between the water above and the water below.

Because the narrative of Genesis 1 seems to place the emphasis on the formation of the Earth, the Moon, and the Sun, human beings for thousands of years have been having a hard time explaining Genesis 1 using the current scientific knowledge which has provided information far beyond the visible night sky. Because, in the days of Moses, all that people could see was the night sky; his story may be relating to the things visible to the naked eye in those days, which is significantly less than what current telescopes can see. It is possible that Moses focused part of his story on that limited sky, which is not even all the sky in the Milky Way Galaxy, the galaxy that the Sun belongs to.

However, the creation story in the Book of Enoch detailed things that were precursors of the universe in a language that even the most advanced contemporary scientists may not understand. For those revelations are about things that scientists may never physically see, for they no longer exist, and no physical terms or experience could allow them to properly describe them without relying on belief. In my book called *"Origin of the Spiritual World,"* I elaborated more on those mysteries.

As God was separating the waters from waters on Day 2, He could have ensured that a proper amount was left below the sky, or else the size of the Earth could have been different and many environmental factors could have differed in such a way that life could have been different. For a bigger Earth or a smaller Earth could offer different conditions for life. In other words, God's design and leadership during the formation of the Earth played a key role in ensuring everything was done to meet His plan. As these ideas came into my mind on December 25, 2013, I did not even know that a day would be coming 12 years later when I would be writing about the origin of the universe.

On May 10, 2017, months before I discovered turbulence in the data I was analyzing, it appeared to me that the division and aggregation of waters mentioned in Genesis 1 must have been very fast and that similar processes could have occurred simultaneously across many other precursors of systems of bodies in the universe. In other words, I felt like just as the precursor of the Earth was undergoing changes, that of many other bodies could have been doing the same as well according to the (split-gathering) laws I was "struggling" to scientifically demonstrate in those days.

I remembered that in early 2017, I used to take water in a bowl and throw it in the air just to see which physical mechanism could illustrate or explain how water breaks up. If you take water in a big bowl and then start moving it, you will see that the water will start moving, and according to the force you are using, that water can start breaking and even escape the bowl before you intentionally throw it in the air.

As I was doing this experiment around mid-2017, I noticed that the initial volume of the water that I threw in the air was split over many smaller volumes of various sizes. During those experiments, I noticed that some blocks or drops of water were small, while others were big. In those days, although I did not yet know the scientific processes by which the water from the bowl was breaking apart, I was amazed at how the creation story narrated by Moses reflects realities according to which an initial amount of highly moved water could have been broken up and separated into different layers and created a space or sky between them.

In those days, I already understood that the direction into which the water of the precursors was moved could have impacted the direction or sense of the revolution and rotation of the Earth. For the precursor of the Earth moved in the direction it was launched. Considering the orbital speed and rotational speed of the Earth, I used to suspect that the waters of the precursor of the bodies in the Solar System could have moved at a very high speed, which also aided in quickly separating the waters of the precursor of the Earth. Because the Bible said that the Spirit of God moved at the surface of the water since the first moment of the first day of creation, and because it was on the second day that the separation or breakup of the waters of the precursors of the Earth was reported, I deduced that it could have taken 1–2 days maximum for the breakup of waters related to the precursor of the Earth to happen. Although these timings may sound trivial for some people, they were a key I used to guide myself until I unraveled the high-level mathematics encrypted behind what Moses said thousands of years before science was even invented.

In the days that I was first inspired to illustrate fluid separation using moving water in a bowl in 2017, I used to show someone near me many times. That was an "aha moment" for me and probably for that person, although I was not sure that person could fathom why I was so thrilled at trying to understand a simple creation story that billions of human beings before us have read and believe or deny. I even kept a frozen bowl of water from some of my "aha moments" in my freezer since 2013 and instructed people in my house not to throw it out. Until 2025, that 12 years frozen water is still in my freezer! Yes, even years before I discovered turbulence, I was mysteriously driven to deal with water in a special way.

Around mid-2017, I tried very hard to figure out the scientific phenomenon that can explain water separation. Because I was not trained as a physicist during my graduate study in the USA but as a plant scientist, an entomologist, and a microbiologist, and all of the math and physics I used in my books was what I learned in Africa before moving to the USA, it took me months of reading scientific articles before I realized, using my high school physics knowledge, that some properties of water separation that I was looking for were handled in fluid dynamics. I had to read hundreds of articles about fluid splashing, sloshing, and other aspects of fluid breakup before I realized that a phenomenon called "turbulence" could have been in play. After that discovery, it took me almost 3 more years working day and night on the scientific data before I finally understood what happened during the formation of the universe: a gigantic turbulence took place and shaped everything…. In my book "Turbulent Origin of the Universe," I wrote more than 500 pages to address that topic.

6.6. Are the creation days really 24 hours each or millions of years long?

Do many Christians deny God's version of the origin of the universe, yet they accuse unbelievers of doing something they themselves delight in doing but don't know?

The birthdates of the Earth, the Moon, and the Sun proved that there is no gap between them or before them and that the days are really 24 hours each. Indeed, without a doubt, the scientific evidence (that I extensively explained for the first time in history) showed that the creation days were 24 hours each. This proves that all anti-creationist theories and even some forms of creationism (particularly all versions of Old Earth creationism) are plainly wrong. Although "Young Earth creationism" defends that God created the Earth in six 24-hour days, none of its adepts has laid out any comprehensive scientific demonstration of how the Earth, the Moon, and the Sun were formed by the 4th day of creation. As I carefully reviewed their arguments against other creation narratives and anti-creationist ideologies such as evolutionist theories, I came to realize that even the proponents of Young Earth creationism have been making mistakes in their explanations of creation. In other words, it would be wrong for me to say that everything that the supporters of "Young Earth creationism" said was true. Nevertheless, that group of creationists was the closest to the truth, but they just failed to scientifically demonstrate it.

To distance myself from certain errors in all existing creationist theories, I, Nathanael-Israel Israel, coined the term "Science180 creationism" to label how I properly explained creation biblically and scientifically. In other words, Science180 creationism is the umbrella under which I put everything I demonstrated in this book and in the 8 others that I wrote on the creation of the universe with emphasis on the God of Israel. For, after all, the Earth is neither old nor young, for nothing is older than eternity and nothing is younger than now. People die at the age of 120 years, and they are said to have been old. Compared to eternity, a few thousand years is nothing. Hence, the Earth, which is less than 6000 years old (as I demonstrated in my book on the age of the universe), cannot be called young nor old. With the extensive demonstration I did to prove the accuracy of the six 24-hour days of the creation story of the Bible's Book of Genesis, the debate between Old Earth creationism and Young Earth creationism should be considered closed, and the answer is Science180 creationism. See more details in *Reconciling Science and Creation Accurately.*" Anyone who will keep arguing about the age or the formation of the Earth being millions of years, is just entertaining a heretical nonsense. The universe was NOT created or formed after billions of years of processes and no book other than the Bible properly narrates creation and its purpose and fate. There is no gap between or before the days of creation. I proved that the formation of the Earth was indeed completed on Day 3, while that of the Moon and the Sun was on Day 4. There is no gap and this gap nonsense that opened heretic doors for evolutionism to enter the church and sit even in the pulpit needs to stop now.

Why then, for a long time, have some Christians argued that the creation days are not 24 hours each? These kinds of arguments caused problems for rationalists who know the flaws of some creationist "demonstrations," and therefore they mistakenly throw away the entire Bible. Please see the detailed answers in my book *"Reconciling Science and Creation Accurately."*

6.7. Were all celestial bodies really formed completely by the end of the 6th day of creation

Have you ever wondered if everything in the universe was really formed in 6 days? Well, I want to share with you a revolutionary approach that will help you. The math and science I did to prove the Biblical account of creation also proved something we cannot deny.

Just as human beings are born as babies but have to go through some processes before becoming adults, so also, in the beginning, the precursors of the heavens and earth were created, but they had to go through changes before becoming the heavens and earth we know today. The Bible even says that, at one point during its formation, the Earth was void and formless, suggesting that even the precursor of the Earth itself passed through some stages of completion. But the problem is that people did not know that all these processes lasted just a few days. In my book *"Reconciling Science and Creation Accurately,"* I dug into the details about the Genesis story.

About a century after Isaac Newton (1642–1727) published his groundbreaking work on the theory of gravity, the British theologian Adam Clarke (1762–1832) referred to this state of the early universe as a *"vast collection of indescribably confused materials, of nameless entities strangely mixed, a crude and indigested state of the original* substance … *that God spent 6 days assimilating, assorting, and arranging into all of the bodies in the* universe[18]." As that theologian put it more than 2 centuries ago, "the heavens and the earth" mentioned in the first verse of the Genesis story, and which derived from the Hebrew words "eth hashshamayim veeth haarets," can be rendered as *"the being or substance of the heaven, and the substance of the* earth." In other words, what Moses was recounting in Genesis 1:1 was not a finished or complete heaven and earth, as some people think, but their precursors, which, after going through some processes, yielded them. The precursor that I referred to in my writing as "turbulent prima materia" is the materials, the supplies, and the ingredients that God created in the beginning and out of which He built the whole universe. Therefore, the sentence "God in the beginning created the heavens and the earth" has been well understood and rendered by some theologians many centuries ago as *"In the beginning, God created the substance of the heavens and the substance of the* earth." In other words, just as your date of birth is not the date you acquired all the form, shape, maturity, and knowledge that you have today, so also celestial bodies were born within the 6 days of creation, but their precursors had to go through additional processes so their formation could be completed. However, nobody before me ever knew how to scientifically prove that until my discovery, which is acknowledged as the milestone that truly reconciled science and the Biblical creation.

As I detailed in my book *"Turbulent Origin of the Universe,"* many celestial bodies were formed even after the end of the 6th day of creation, but their precursors were in place and were going through the changes needed for these bodies to be formed before the end of the 6th day. In other words, when the Bible says that God created everything in 6 days, it does not mean that the formation of all the celestial bodies in the universe as we see today was completed by the end of the 6th day. In the aforementioned book, I also proved that, even in the Solar System, some planets were not completely formed before the end of the 6th day, but their babies or precursors were already formed, and just like a child must go through some developmental stages before becoming an adult, the precursors of some celestial bodies had to go through some processes before becoming what they are today. However, based on my math and science, those processes for the planets in the Solar System were a matter of days, not even months. Just like a pregnant woman can give birth to a child after about 9 months of pregnancy, and that child must go through stages before becoming an adult, so also the precursors of the celestial bodies were all born by the end of the 6th day of creation, and some of them had to continue their journey in space before reaching the position where their formation was completed, and they were put in orbit.

Furthermore, if you study the meaning of the Hebrew word translated as "heavens" in the Genesis story, you may be tempted to think that Moses was not talking about all of the universe. I know some people, even some Christians, will want to fight me for saying that God did not form everything in 6 days. What people don't understand is that understand, there is a huge difference between creation and formation. Yes, God created everything in 6 days, but everything was not fully formed before the end of the 6th day of creation. By the time God was resting on the 7th day of creation, as the Bible says, I can assure you that the that, the math and science that I used to prove the Biblical creation also say that some celestial bodies were still being formed. As I proved in my book *"Turbulent Origin of the Universe,"* Jupiter was not even formed when God finished His rest on the 7th day. In that book, I explained the exact date that all of the planets in the Solar System were formed. Maybe, you are one of the people who may ask, "Nathanael-Israel Israel, how could you say this? Come on, Nathanael-Israel Israel, give me a break! Give God and the Bible a break! That sounds so crazy."

Well, I understand why you would say that, but here is the fact: the math and science that proved that the Biblical creation is correct also say that some celestial bodies were still being formed by the end of the 7th day after God rested. Because I already argued that thorny remark in my book *"Reconciling Science and Creation Accurately,"* I will not debate it here again. But if you want to know more about it, you can refer to that book.

In that book and in *"Turbulent Origin of the Universe,"* I scientifically proved that the beginning of the solar system is also the same thing as the beginning of the whole universe. Therefore, even if some people may feel more comfortable considering the heavens mentioned in the first verse of Genesis 1 as the Milky Way Galaxy or even the Solar System, still the Genesis story can be applied to the beginning of the entire universe. And because all stars and celestial bodies are not the same, some could be born sooner or later than the Sun. But the reference to the beginning of the universe could still be connected to the story mentioned in the Biblical account of creation.

As you can see, there is a lot more to say than what I can summarize in this book. Hence, I urge you to get my other books (see www.Science180.com/books) to familiarize yourself with other aspects of the formation of the universe I could not handle in this book. That way, you can properly answer other questions you have and better equip yourself to handle other people you may meet along the way.

Because they failed to understand the process that God used so that the Earth could be formed by the 3rd day of creation and the Sun and the Moon on the 4th day, some believers who dearly hold onto the Bible story mistakenly think that God created all of the other celestial bodies (e.g., galaxies, stars, asteroids, planets, satellites) in the universe by the 4th day, implying that they even think that Mercury and Venus were created on the 4th day, not knowing that these 2 planets were even created before the Earth (see my book *"Turbulent Origin of the Universe"*). For these groups of believers, every celestial body in the universe must have been created by the end of the fourth day of creation. However, my discovery showed that, although some planets (e.g., Mercury and Venus) were formed before the Earth, and that besides Mars, which was formed on Day 5 (see the book I just mentioned), all other planets in the Solar System were not even formed before Day 7, when God was resting. Likewise, although several celestial bodies including stars in the universe beyond the Solar System, could have been completely formed by Day 4 and even by the end of Day 7, several others were formed afterwards. By the end of Day 7, the process of the formation of most celestial bodies that were not formed yet was launched, and nothing could have changed it. In other words, by the end of the 7th day of creation, everything we see today was already in a form of existence. When God was resting on the 7th day, He knew that everything not formed yet will certainly be in time according to its their order. In other words, it is indeed correct to say that God finished creating things by Day 7, but the formation of several things continued afterwards. Even certain things are being created today, but not as they were done in the beginning.

Every birth of a living thing is a miracle of creation, although some people may just see it as a process of reproduction. If an entire human being can be conceived, carried in a mother's womb, and born within a few months, why do we need to think that it should take millions of years for a human being to be formed for the first time or that it should take the same amount of time for the universe to be formed? Some of those who believe in God can testify to having witnessed miracles done "by" some preachers who, after speaking or declaring the word of God, made supernatural things happen, including creating things, healing incurable diseases, and performing signs and wonders beyond the ability of a mere human being. For God can empower believers to do mighty things in His name and for His own glory! And you want to tell me the universe was not created in 6 days?

6.8. Why is water confirmed on most celestial bodies in the Solar System?

As of 2025, scientists at NASA and at other organizations have confirmed the presence of water on many celestial bodies, including the Sun and all of the planets in the Solar System. According to the Biblical story of creation, water was indeed present from the second day, suggesting that it could have been formed by then. Because they failed to understand the origin of the chemical elements and of the celestial bodies, some people (even scientists) wrongly believe that water was brought onto the Earth by a comet that collided with our planet. Yet, even recently, water was again confirmed on our neighboring planet, Mars.

For instance, because the fluid layers of the precursor of Mars was at one point close to the fluid layers of the precursor of the Earth-Moon system, its position favored the formation of a wetter climate early during its lifespan. Although by the time the precursor of the Martian planetary system reached its position, the environment could have not allowed water to be formed abundantly, some water could have been formed on it during the journey from its separation from the Earth-Moon system until it reached about the orbit of Mars, which is about where the precursor of the Martian planetary system split into the precursor of Mars and the precursor of the Martian satellites. I handled those aspects in my book on the origin of the chemical particles: *"Turbulent Origin of Chemical Particles."* Because some believers do not understand that during their formation the fluid layers of the Earth-Moon system and the Martian planetary system were next to each other and had the precursor of many chemicals in common before each of them finished their formation, they tend not to be able to explain water on Mars by any other means than the Great Flood from Noah's time, which is not necessary to explain the matter. In other words, those believers think that the water found on Mars and on other planets in the Solar System must have been brought there by the water of the Great Flood during the time of Noah. To put it another way, they think that the Great Flood of Noah's time occurred not only on Earth but also on all of the planets in the Solar System as if God had put a huge belt of water around the Solar System or around some planets of the Solar System and then unleashed that water during the flood. As for me, it is the misunderstanding of the Biblical story of creation and the wrong theories some believers have forged that caused them to have such a wrong perspective about what happened in the beginning.

- Why then are scientists exploring water presence on other celestial bodies? It is a waste of time and resources, for they don't know how the Solar System was formed...

- Why explore Mars and the Moon for water presence? They won't find many new things, for Mars, the Earth, and the Moon are like brothers, sisters, and cousins, and if we understand the Earth, you won't need to waste our time on any of these celestial bodies. Maybe some people should consult with me (really) so I can help them not to waste part of the money saved from government efficiency on useless Mars explorations that are being championed.

- Do I know how the universe works better than some chainsaws that cut governmental spending in recent years?
- Why then are we wasting a big chunk of the governmental money on some useless projects at some agencies or organizations seeking to understand the origin of the universe, of chemicals, and of life beyond the universal scope I laid out in my books?

6.9. Why is the Islamic creation narrative different from the Biblical account of creation?

I cannot reach this point without saying why the Islamic creation narrative is different from the Biblical creation story, yet they are all called creationist stories. I started addressing this issue earlier in Chapter 5 on the world's main religions, but I had to pause the discussion because I felt like I needed to address certain issues first. Now, I will elaborate more on this challenging issue I cannot ignore.

Although some people may quickly blame the author of the Koran for the big mistakes in the Islamic account of creation (for it contradicts the Biblical story, which I proved to be scientifically correct), I would like to place the emphasis on how, although the Jews and Arabs share a common ancestor, the way the creation story may have been transmitted orally throughout generations may have falsified the Islamic account of creation.

The ancestors of Islam may have learned part of their creation story from Abraham or his descendants. To my understanding of the Bible and history, both the Jews and the Arabs descend from Abraham; hence, Islam and Judaism are called Abrahamic religions. How did they end up having different stories of the formation or the creation of the universe by God?

Considering the animosity between the Jews and the Arabs, it is possible that the founder of Islam may have been misled by a demon who portrayed himself as Angel Gabriel, who is mentioned in the Bible as a good messenger. Either that demon misled Mohammed, or Mohammed himself or his followers (who wrote the Quran) invented the story of creation using what he knew from the Jewish story of creation, which was known for almost 2000 years before his days.

I don't think that Mohammed intentionally lied. I think he might have encountered an angel, but not a holy angel from the God of Israel. For God cannot contradict Himself.

Considering the scientific evidence I presented in this book and others I wrote, I strongly believe that the Koran is NOT a revelation from God that the Muslims call Allah. The math and science that I showed in my books overwhelmingly proved that the Koran is a corrupted message from a demon that Mohammed mistook for Angel Gabriel, the angel that God used to announce the birth of Jesus to the Virgin Mary.

The same holy angel Gabriel cannot reveal Jesus to Virgin Mary as the Messiah of the whole world, and then 600 years revealed to Mohammed that Jesus is just a simple prophet, not even the Savior. Being a God of order, the God of Israel who shared the Biblical creation story with Moses could not have shared a contradictory story of creation with Mohammed. Considering the historic animosity between the Jews and the Arabs, and considering that the scientific evidence confirms the Biblical account of creation, I cannot draw any other conclusion on the Koran than what I said above: it is a corrupted book that tried to paint the Jews as the enemies of God. But are all Jews children of God? No, of course; and many of them will go to hell just as all Muslims who will not believe in God before the Judgement Day. The same applies to anyone else (regardless of their religion) who denies God until the Judgement Day. Although some people may hate the following, I also believe that some people who ignore or deny God today while alive may get another chance even after their death, for, considering what the Bible says about Jesus Christ preaching to some dead people during his 3-day days stay in the grave, it seems like some dead people, at least those in the Days of Noah, may have gotten a second chance to hear the Gospel from Jesus and get another opportunity to choose between believing in God or denying Him. Because that is true for those people, can we be correct by saying that some unbelievers today may not get another chance in this life or after their death, before the Judgment Day? This possibility is not an excuse nor an encouragement for anyone to deny God today, but a warning that everyone should be working on his or her own salvation, for we don't know what is causing some unbelievers to deny God today, and we don't know for sure if any of them may get another chance to believe or not someday, even after their death. For instance, I cannot close this segment without reporting what some prominent prophets strangely said recently: that Judas Iscariot, who sold Jesus Christ and who committed suicide, is not in hell but, but in heaven. If that is true, and I won't be surprised if it is, it means that Judas Iscariot may have gotten another chance to repent before his death during his agony or even after his death. Who can verify that for sure? Not me! Who can say it is impossible? Not me either! Therefore, let's be careful about how to view some unbelievers today, for some of them may end up making it to heaven after some experience before their death, or after a second chance they may get even after death that some believers won't get. God is the Judge, and only He knows for sure the final criteria of grace. Let us not judge, and let us show mercy to others. Still, this precaution is not an excuse to avoid telling people the truth.

I cannot, and I don't want to, finish a book about the creation of the universe without clearly drawing the lines between the Biblical account of creation and the Islamic statement on creation. The scientific and mathematical demonstrations I did in this book and others I published in 2025 and beyond proved that the Koran is not God's own divine speech, as Muslims claim. In fact, even Muslim scholars agree that the Koran itself was not written by Mohammed, or that he did not write down the revelation himself, but his disciples or followers wrote it.

Furthermore, Chinese, Japanese, Nepalis, and Indians descended from lines or lineages existing before Moses was born. They don't have Abraham as their ancestor, and they were probably unaware of the ancient or sacred story of creation, which the ancestors of the Abrahamic religions knew. Hence, the eastern religions (Buddhism, Hinduism, Confucianism, etc.) born in those regions have no clue about the creation story. Most of these religions did not come close to the Biblical account of creation, which is the only story that is backed by science. Buddhists don't even talk about creation. I think their ancestor and his followers were frank not to invent a creation story and force the Buddhists to embrace it. In contrast, Hindus have many versions of creation and many beliefs, but they are all completely different from the Genesis narrative in the Bible. Animists in Africa also have many versions of the world's origin, but they all are absurd.

Therefore, I realized that the creation stories of the world's religions boil down to what their founders or ancestors knew or did not know about creation from Moses' narrative! Although Moses existed around 1500 BC, meaning that the Torah was written around that time, Jews may have already known about creation through the works of Enoch, who was the 7th generation from Adam. In my book *"Origin of the Spiritual World,"* I elaborated on the depth and accuracy of many scientific things revealed in the book of Enoch, which is even older than the Bible. Some lessons and stories that Enoch and Moses taught may have been passed onto the forebears of the Arabs, for both the Jews and the ancestors of the Arabs were all living in the same area, the Middle East. Therefore, it should not be surprising that the Islamic account of creation also talks about 6 days of creation, although it does not match the details of the Biblical account of creation. In fact, the Koran does not start with or have a clear story of creation, but creation references were made here and there in that religious manuscript. Mohammed, the founder of Islam, died around AD 632, meaning about 600 years after the death of Jesus, the main character of the Bible. By then, it cannot be excluded that the ancestors of Islam knew of and may have even read the Genesis story in the Bible. It is possible that the Muslims may have copied part of their creation references from the Bible and/or reported what they remembered from what they were told centuries ago and/or twisted it to make their case against the life of Isaac and his son Jacob, the patriarch after whom the name Israel was given.

6.10. Take-home message

As I scientifically proved in this chapter and in the last 2, the creation story in the Bible's Book of Genesis perfectly matches the scientific evidence concerning the formation of the universe, according to which Moses mentioned that, on the first day of creation, God created the heaven and Earth, but on that day, the Earth was still not fully formed yet, and it was on the 3rd day that the formation of the Earth was completed, while it was on the 4th day that the formation of the Moon and the Sun was completed.

Indeed, in the beginning of the universe, precursors of bodies were formed, fluids layers were formed, separated, and gathered together. The separation of the fluid layers of the bodies orbiting the Sun started no later than on the 2nd day. The formation of the Earth was completed on the 3rd day, while that of the Moon and the Sun was completed on the 4th day, just like the Bible says. As the next chapter of this book will clarify, the original Hebrew texts of the Book of Genesis clearly explained that the Earth mentioned in the first chapter of Genesis was the precursor of the Earth, which had to go through some changes before birthing the Earth we know today. It is the English translators of the Bible that made it sound as if the Bible says that the Earth was fully formed on the first day.

Unlike what the Bible correctly revealed about 3500 years ago, the Koran wrongly said that the Earth was created on the 1st day and that the Sun and the Moon were created on the 6th day. In light of all the evidence I presented, the creation story promoted by Islam in the Koran does not match the scientific evidence, which is perfectly aligned with the Biblical narrative.

No other creation story of any other religion in the world has ever said anything matching the Biblical narrative of creation that I scientifically proved as the only true story of the formation of the world. None of the scientific theories ever built matches the Biblical narrative, as I demonstrated in this book and others I have written on the origin of the universe and everything in it. The problem of the delay in proving the Biblical creation was not the lack of data, but was about how to properly analyze the data! Voilà!

- Why did it take so long, more than 3500 years, before the Biblical account of creation was finally scientifically demonstrated? Why now? What is its significance, and what is going to happen next?
- Why did the ancient Greek thinkers born about a thousand years after Moses miss the formula that leads to God's existence?
- How and why did Moses, who lived thousands of years before the scientific age, accurately predict the Biblical creation narrative?

If you don't want to miss the shocking answers to these questions, then flip the page to the next chapter.

7. HISTORY AND SIGNIFICANCE OF THE PERFECT MATCH BETWEEN SCIENCE AND A RELIGIOUS STORY OF CREATION OF CELESTIAL BODIES

Before I handle the significance of the accuracy of the Biblical creation narrative, I would like to first explain a little bit about the historical context and the stage of scientific knowledge in the days of Moses and afterwards.

7.1. Why did the ancient Greek thinkers born about a thousand years after Moses miss the formula that leads to God's existence?

According to historical evidence in the Bible and in secular references (more details in my incoming book on the age of the universe), Moses could have been born a little before 1500 BC, for he was about 80 years old before Exodus. In my book on the age of the universe, I elaborated on the timeline of other key events in the universe.

The creation narrative in the Bible's Book of Genesis was recounted about 1000 years before the Greek philosophers regarded as the ancient fathers of most scientific disciplines were born:

- Homer (born c. 750 BC)
- Thales of Miletus (c. 624 - c. 545 BC)
- Pythagoras of Samos (c. 570 - c. 495 BC)
- Heraclitus of Ephesus (c. 535 - c. 475 BC)
- Anaxagoras (c. 500 - c. 428 BC)
- Protagoras (c. 490 - c. 420 BC)
- Socrates (c. 470 - 399 BC)
- Hippocrates of Chios (c. 470 - c. 410 BC)
- Democritus (c. 460 - c. 370 BC)
- Leucippus (460 - 4th century BC)
- Plato (428 - 347 BC)
- Diogenes (c. 412 - 323 BC)

- Xenocrates (c. 396 - 313 BC)
- Aristotle (384 - 322 BC)
- Euclid (mid-4th century BC - mid-3rd century BC)
- Epicurus (341 - 270 BC)
- Aristarchus of Samos (c. 310 - c. 230 BC)
- Archimedes (c. 287 - c. 212 BC)
- Eratosthenes of Cyrene (in modern Libya) (c. 276 - c. 194 BC)
- Diocles (c. 240 - c. 180 BC)
- Hipparchus of Nicaea (c. 190 - c. 120 BC), the founder of trigonometry

In the days of most of these Greek philosophers, mathematicians, astronomers, and various thinkers, most of the basic things taught to children in elementary school today were not even known yet. The kind of knowledge that most elementary students in the 21st century would consider common sense was what the "top" thinkers or pundits who lived 1000 years after Moses would have called advanced "scientific discoveries." If anyone had told the Greeks who lived 3000 years ago that human beings could be able to do what we are doing today with technology in the digital world, they would have vehemently said NO. Nobody who lived a few thousand years ago could have imagined that we would have all the privileges we are enjoying today in the 21st century. In other words, very little was known about nature in the days of those Greek pundits, and much less in the days of Moses. It was progressively that, through the pioneering work of many Greeks and others, the foundation of what would later be called science was laid.

Indeed, Greeks have significantly contributed to founding many Western scientific fields, notably astronomy, mathematics, and much more. When the Greeks, or Hellenes, are mentioned, many people tend to think that they are only Europeans or Asians, but, in fact, some Greeks were Africans (e.g., Egyptians, Libyans, etc.) who made significant contributions to the advancement of human knowledge.

Now, I will illustrate why I said that what the most "educated" people knew more than 1000 years after Moses had died was very rudimentary. For instance, although the distance to the Sun and the size of the Sun have been speculated since the days of Aristarchus of Samos (c. 310 - c. 230 BC), a Greek, it was in the 20th century that their "precise" values were established.

Likewise, the radius of the Earth has been estimated to a reasonable accuracy for more than 2000 years by an African, Eratosthenes (c. 276 - c. 194 BC) of Cyrene (located in modern Libya). In other words, that African was the first person to calculate the circumference and the axial tilt of the Earth. More than 2000 years after the days of Moses, human beings were not even able to know whether the Earth orbits the Sun or the opposite yet. For instance, in the second century after the birth of Jesus Christ, the so-called best theory that the ancient astronomers produced was a geocentrism theory according to which the Earth is at the center of the universe.

Designed by the Greek scholar Claudius Ptolemy (AD c. 100 - c. 170), that geocentrism theory prevailed for more than 1500 years. In other words, even 1500 years after Jesus Christ, meaning more than 3000 years after Moses recounted the Bible's story of Genesis, the most learned people still believed that Venus, Mercury, and the Sun orbit the Earth. It was around the death of the Polish mathematician and astronomer Nicolaus Copernicus (1473-1543) that it was accepted that the planets (including Earth) orbit the Sun. Before then, some people, even prominent ones, accepted that the Sun was the center of the universe: this was the heliocentric system. Even so, many people still believe that the Earth is the center of the universe. When the Italian astronomer Galileo Galilei (1564–1642) was saying that the Earth was not at the center of the universe, he was not believed, and he was jailed and then killed. A few years later, the work of the German astronomer and mathematician Johannes Kepler (1571–1630) confirmed the hypothesis evoked by Galileo. Until his invention and/or "modernization" of the telescope in 1610, no one had ever observed a celestial body (other than the Moon) orbiting a planet.

In 1619, Johannes Kepler proposed his 3 Laws of Planetary Motion. One of Kepler's laws was about how the semi -major axis and the orbital period of the planets are related. That is the 3rd law of Kepler, or the law of harmonics, applied until today. This means that since the days of Kepler (more than 400 years ago), the semi-major axes axis of some key planets were known with an acceptable accuracy. I hope you remember what "semi-major axis" means: the average distance separating a body from its primary body. For instance, I told you earlier that the semi -major axis of the Earth is the average distance separating the Earth from the Sun.

It took more than 45 years after the death of Galileo before the English mathematician, physicist, astronomer, and theologian Isaac Newton (1642–1727) elaborated (in 1687) the so-called "laws of motion" and "universal law of gravity" that dominated science for centuries until the days of Albert Einstein (1879–1955), the German-born physicist who elaborated the influential theory of relativity and who is perceived by many as the greatest physicist of all time, just next to Isaac Newton. But is that really true?

The so-called universal law of gravitation (elaborated by Isaac Newton) speculated that "all objects in the universe attract all other objects, and that the attractive force is related to the mass of the two objects and their distance apart." I deliberately used the words "so-called laws of motion" and "so-called universal law of gravitation" because, as I proved in my book *"Turbulent Origin of the* Universe," these laws don't properly justify gravity. I dealt with this law (and its weaknesses) in the chapter on gravity and the fundamental force in nature in my book I just mentioned.

Moreover, the escape velocity that I used to estimate the timeline of the formation of the celestial bodies was derived from the work of Newton, meaning that since the days of Newton (about 300 years ago), people could have ting the mainstream scientific priorities and some religious dogmas dictate what they should think about, how they should think, and what they should do before getting a salary or a promotion, provided they were ready to sacrifice themselves for others as I did for the last 12 years. More details about my journey and sacrifice can be found in my incoming autobiography.

Many details about the solar system were known just recently, particularly in the 20th century, the decade during which human beings were able to go into space for the first time. As more data were collected by scientists, many that they supported for years came to be wrong, and adjustments have been made even until now. Yet, despite the massive data they have collected and that they keep collecting on the universe, scientists have not been able to use them to properly explain the origin of the universe or to properly decrypt the message left by Moses more than 3000 years ago, and some scientists have regarded as unscientific, false, fictive, and unworthy to be believed or investigated. Yet, I did not need to edit or change anything in the creation story of Moses before fitting the scientific data with it. What an error people have made concerning the creation story and many other stories in the Bible!

Based on the quick history I just provided, I am not the one who discovered the radius, the semi-major axis, the escape velocity, and the orbital speed of the celestial bodies, but I am honored to be acknowledged as the first person in history to discover how to use these variables to properly explain the origin of the universe and the existence of God. I am fortunate to be the one who created the formula to calculate the split-gathering and the birthdate of the celestial bodies. I am privileged to be the first person to offer a comprehensive explanation of turbulence, a scientific field not well understood until I explained it, and without which it was hard to comprehend the formation of the universe. For more details on turbulence, see my book *"Turbulence Origin of the Universe."* You can also visit Science180Academy.com to apply for "Science180 Academy of Turbulence," where I help turbulence experts to get their turbulence questions answered.

However, considering how I reached this level of achievement, I realized that, if people had properly aligned their thoughts with the Word of God, without which I could not have navigated the countless obstacles I met during the 12 years of deciphering the code of the universe (which I could not even fully present here, hence I wrote other books about it), they could have proven a long time ago that the Bible told the truth about the origin of the universe. Yet, countless scientists have devoted their lives and careers to spending trillions of dollars poured into "scientific" research for centuries (at the expense of crucial needs to achieve better results) to explain in vain what I did in a few years with zero funding for my research from "anybody," except from God, who has been providing for me supernaturally for these 12 years during which I had no salary and no (significant) income coming in my name, and people were mocking me and calling me by several unapplicable names, not knowing that my achievements and my God would openly speak for me one day. To know more about this, please contact me at Israel120.com/contact.

In my incoming autobiography, I explained the sacrifice I had to make for my career and life before reaching this stage. I thank God for the process, for if people and organizations had known what I was investigating about the universe's origin and had invested in it, they would have already pushed me a long time ago to produce precipitated reports to justify their investments and the usage of my time, therefore forcing me to draw unthoughtful conclusions too soon before I could even "holistically" figure out the best way to tell the story of the formation of the universe to all kinds of people. I worked on this project for 9 years, day by day, before I finally discovered the formula of the birthdate of the celestial bodies, without which it could have been impossible to convince some people (even the creationists) to accept that God created the universe in six 24-hour days just as the Bible revealed more than 3000 years ago. After those initial 9 years, it took me additional 3 years before I went live with my findings in 2025. Therefore, you need to take the findings in this book very seriously, for people, including myself, have paid a huge price, and others have wanted to learn about what you are graced to read in my books, but they did not have that chance! The next segment will illustrate what I just said and how it can change your life.

7.2. How and why did Moses, who lived thousands of years before the scientific age, accurately predict the Biblical creation narrative?

Despite having lived about 3500 years ago, meaning more than a millennium before the days that science was invented by human beings, Moses was so accurate. Did God ordain a holy angel to reveal to him or did God write the story and hand it to him just as God did for the tablets of the 10 commandments? Besides Jesus, Moses was the greatest prophet of the Tanakh (what the Christians call the Old Testament), even the greatest prophet of all times!

Moses knew a lot of things that people throughout the ages have failed to understand. Based on the demonstrations I did and which confirmed the story that Moses recounted about the first 4 days of creation, I now know that the Biblical creation story is the most misunderstood and least dissected story in the Bible and even in the whole world throughout the ages. In other words, Moses was misunderstood by so many including his own people and most of the other people who lived on this Earth (even those who believe in the Bible). Put another way, in addition to all kinds of unbelievers who purely rejected God and the creation story, all (except a handful of people) believers did not grasp the fullness of the creation story. Else, what I demonstrated in this book could have been scientifically established at least a few centuries ago. Now that I proved it, will even believers embrace it?

I have read many Biblical commentaries written by theologians throughout the ages, but I have not found a single one who understood the entire creation story in the Bible; for some get some parts right and other parts wrong. Most people got the entire story wrong and landed their logic into a billion-year process theory, including evolutionism, yet some of them call themselves Bible believers. Moses also knew that the children of Israel (Jews) he was talking to have no clue of what God showed him.

This could explain why Moses was so angry with them on many occasions, to the point that God had to punish his anger and prevented him from entering the Promised Land. Here is the man whom God used to split the Red Sea, the man who, for 40 years, prayed, and foods, including bread (manna) and meat (roasted quails), fell from the sky so that the Israelites could eat during their deliverance journey from Egypt to the Promised Land. This man prayed, and water came out of the Rock, the Rock of Ages, which Apostle Paul later referred to in the New Testament as Jesus Christ following the Israelites during their journey in the wilderness. This prophet Moses, whom God used to give the law to His people, was also one of the first to break it, including marrying a gentile woman, an act that goes against God's commandment, which forbade the Israelites from marrying strangers. Yet, by His grace, God still saw him favorably.

When I considered everything, I felt like all deep mysteries are not meant to be understood physically by everyone, much less with a mere mind, but to be believed deeply so that the unquestionable belief can supersede any doubt the mind, heart, and brain may have, in such a way that the believer can be perfected not on the basis of physical knowledge, but on the basis of spiritual truth believed, even if it is not understood. That is faith! And that faith can discard useless hypotheses and questioning and lead to the path of scientific proof, even when you don't know science or are not trained as a scientist! Hence, for millennia, people have read, memorized, and even taught the Biblical story of creation put as the first story in any Bible, in plain sight, but very few individuals have ever understood it or known what Moses meant or even entertained the thought of investigating it further. Besides God the Creator, Moses, and Enoch (the grandfather of Noah), I wonder how many other human beings who lived on Earth have really fully understood what the Bible says about the first 4 days of creation. In my book *"Turbulent Origin of* Life," I scientifically dealt with how all forms of life were formed, including the events that the Bible says happened on Day 5 and Day 6 about the creation of life! In fact, written in a scientific language that laypeople can understand, *"Turbulent Origin of Life"* is what I called the biological or life version of my book on the origin of the universe. It is meant to suit scientists, nonscientists, and all kinds of laypeople, and it decodes the origin of all forms of life so human beings can understand and better live. As of 2025, *"Turbulent Origin of Life"* is my only book devoted to the origin of all forms of life, and it will help you to grasp in simple language what is needed to fully understand the formation of all forms of life. Whether you are a scientist or a layperson, a believer or a skeptic, you cannot afford to ignore the greater, better, faster, simpler, cheaper, easier, and more accurate formula unlocked in this important book that successfully decoded the origin of life. Get *"Turbulence Origin of Life"* today and change lives. Don't wait. Learn more at Science180.com/life.

Truth be told, despite my efforts to decrypt the Genesis code, I don't even consider myself as having gotten it all yet. That is the wisdom of God: putting deep secrets in plain sight for everyone to see but encrypting their understanding and decoding. in such a way that scientific tools alone can never fully apprehend the depth of spiritual treasures without faith in the Creator! Hence, so-called great scientists have existed and tried to unearth the secrets of the universe but could not, because they neglected to consider the spiritual aspects of the keys required to unlock the code of the universe. The glory of God is to hide things of course, but when I studied things in the universe, I realized that God did not really hide every deep thing as we may think, but the problem is human inability to decipher God's encrypted messages and signals. For God is still speaking today, but the problem is that people are not listening; their minds are inclined or turned toward other things, so-called priorities, or interests, while God's priorities, laws, promises, and guidelines are neglected by most people, including some of those who claim to believe in Him.

And because God also does not always punish people right away, nor does He try to prove Himself right before any human attempt to deny Him, those who will go to hell (the eternal dwelling place of unbelievers after the end of this current world) but who seem to be flourishing today think that they are the best, not knowing what is awaiting them. Sadly, some of those who believe in God but who are struggling today (but who will rejoice in heaven one day) do not always know that they are on the right path and that the current suffering is nothing compared to the incoming joy! And when we talk about heaven and hell, some people oppose it, not knowing that it is by ignorance that discussions about eternity (which are supposed to be at the center of everything we do) have been mistakenly relegated! What a mystery!

How can people then listen to what God is saying if they do not want to listen to Him or His Word? That is a story for another day in another book! Unfortunately, while most people refuse to listen to God, they want Him instead to not only listen to their requests but to also solve them all very quickly and to do so all the time, or else they think God does not exist or is not paying attention. That is why some may ask, "How can a good God exist when there are so many problems in the world?" thinking that God's ultimate mission is to solve all of our problems, including those we create while disobeying Him.

Talking about searching for the truth, I have searched in the scientific and philosophical literature, but I have not found a single human being who has ever demonstrated the creation story in the Bible by proving the timeline of the formation of the celestial bodies like I did. In other words, since the days of Moses, when the Bible's Book of Genesis was written, until now, the story of creation has not been decoded by any scientist until the days of Nathanael-Israel Israel. However, the Books of Enoch (which are older than the writing of Moses) also talk about creation. Indeed, the book of Genesis states that Enoch was a pious man who walked with God righteously to the point that God took him to heaven. He was the 7th generation from Adam, and he lived for 365 years. The books of Enoch even recount that Enoch was the one who revealed certain mysteries to Moses on Mount Sinai… More details about this can be found in my book called "Origin of the Spiritual World."

All I said above implies that early from the beginning, God has shared with human beings their origin and the origin of the world He created. But most human beings have failed to decode the message, and many refused to believe in Him literally, ignoring that God is more interested in people believing in His words than in people discovering scientific "knowledge" that they usually misinterpret!

Some people may try to blame God for not having provided more details about creation, but the fact of the matter is that, even if God had revealed more than He did, many people would still refuse to believe. The more details God gives, the more some people (use their ignorance or unbelief to) find arguments to reject Him. For instance, the creation story in Genesis 2 is a detailed version of part of that in Genesis 1, yet some people think that Genesis 1 and Genesis 2 are contradictory. Therefore, they assume that the Bible mentioned 2 different and conflicting creation stories (Genesis 1:1-2:3 and Genesis 2:4-25). Yet, these people believe countless inconsistent scientific theories, which plainly contradict themselves and the Word of God, so they refuse to believe that God used to create everything (John 1:1). For instance, after so many efforts and theories, like I previously said, it was around the middle of the 16th century (around the death of Nicolaus Copernicus) that people "accepted" that the Earth orbits the Sun and not the opposite. In other words, more than 3000 years since Moses wrote the creation story, scientists were still not able to properly demonstrate the movement of the Sun.

The size of the Earth and the escape velocity of the Earth and Sun were known less than 400 years ago. Even if God had told Moses the size of the Earth, of the Moon, and of the Sun; the distance separating the Sun and the Earth; the distance separating the Moon and the Earth; the orbital speed and the escape velocity of the celestial bodies (including the Sun, the Earth, and the Moon); and the formula people can use to figure out why He created the Earth on the 3rd day and the Sun and the Moon on the 4th, and even if God showed people that the fluid layers (not just water) were split on the 2nd day, just as described in Genesis 1, and just as I scientifically demonstrated, some people would still refuse to believe in God until today. Even after my historic demonstrations of the creation story, many people will still refuse to believe in God and His word. Many people will still find arguments to justify their wrong mindset and stay with their faulty logic.

Do you know that until today, some people still believe that the Earth is flat? In fact, many scientists will learn about how I demonstrated the accuracy of the Biblical creation story, yet they will still REFUSE to accept it because of imaginary loopholes those people will create in their own erroneous logic and modeling. In other words, some people claim to be scientists going after facts, but most of them do NOT want the facts to lead them to the Truth, and whenever the facts take them toward the road leading to God, they manage to find other hypotheses to deny Him! For the problems of most human beings are not a lack of knowledge, of facts, or of data (as some claimed in order to continue collecting or forging useless statistics based on incorrect theories worse than some religious dogmas they fight in the name of science, which sometimes seems like a religion) but they usually interpret things the way they want, even though their methodology and persistence in the wrong direction could eventually cost lives forever. Come on, Nathanael-Israel Israel, you may retort! Forever? Yes, but you may say no if you want. At least I accomplished my mission by bringing the truth to your precious attention today!

Therefore, God did not bother proving everything to anybody in detail, but He revealed what needs to be known and then expects us to believe in Him so everything else (including real knowledge and all kinds of needs) can be revealed or given to us if we need it. Therefore, considering the history of human thinking, I felt like, even if God had chosen to detail everything about creation to Moses, Moses would have had to stay on Mount Sinai (where the law was given) for more than 40 days. The Bible recounts that by the time Moses came down from Mount Sinai, where God gave him the tablets of the 10 commandments, many Israelites had already abandoned their faith in God and had built an idol (golden calf) that they started worshipping. Imagine God holding Moses on Mount Sinai for months or years instead of the 40 days so that more details could be given about the beginning of the world, what would the children of Israel (Jews) have done? Most of them may have returned to Egypt just as some people will choose to continue to stay in darkness even after I light a gigantic candle on the true origin of the universe. Likewise, although their descendants physically reached the Promised Land, until today, most of them and their descendants never spiritually reached the spiritual Promised Land, which is what they would get in the age to come if they believed in Yeshua so He could deliver their soul from everything holding them in the spiritual bondage of disbelief. In other words, spiritually speaking, most of the Jews (and the gentiles as well) have been and are still in (a spiritual) Egypt and are not ready to be delivered and moved back to the Promised Land that they think is the physical Israel. What a mystery!

Likewise, for more than 3 thousand years, the Bible has been the only book that properly predicted that the universe was created in 6 days. A small group of Christians have been alone in teaching creation, but most of the human beings who have lived on this Earth refused to believe. Because they rejected the truth, they embraced their own lies. Most people (including Jews and Christians who believe in the Biblical creation story) say it is impossible to scientifically prove that God created the world in 6 days. Most people gave up on the test or effort even before starting it. Because no human being was formed yet and nobody was there when the universe was created by God, most people (even believers) think that nobody can "prove" how the world was created and how things were formed. In other words, they think that no human being can "prove" God or how He created the universe. But didn't I, Nathanael-Israel Israel, just prove it to you in this book and in the 8 others I published in 2025?

For more than a century (since the publication of Darwin's book "*On the Origin of Species*" (the evolutionist "bible" of most unbelievers), many Christians and Jews have blamed proponents of evolution (which I don't believe either, of course) as their #1 enemy, whom they claimed to be against creationism. Yet, by ignorance, some (educated) Christians and Messianic believers and even their leaders have opposed scientific facts that could have helped make the case for Biblical creationism. For instance, what can't I say about the opposition that Nicolaus Copernicus and Galileo (both Christians) faced in front of the church leaders, who, in those days, thought that the Earth was the center of the universe …?

Some people may think that the errors of the church leaders "fighting" some great scientists ended in the days of Galileo. However, even as of today, many church leaders are still in the dark concerning the real explanation of the creation story. Even before I finished working on my books on the turbulent origin of the universe, of life, and of chemical particles, I already felt opposition and even jealousy from the side of some Christian leaders, who not only did not understand the technicality of the creation story, but also wanted to position themselves in front with wrong creation theories they crafted (after prophetically picking up a piece of what I was doing), and their theories blatantly oppose the Bible, such as if the 6 days of creation were millions of years long! Anyways, because my goal is not to expose or attack anyone here, regardless of his or her religion, I will not go deeper into these errors or heresies by contemporary Christian leaders in this book. However, I had to talk a little bit about some of these errors because I needed to emphasize how the wrong interpretation of the creation story of Genesis 1 is not found with non-Christians only, but also even among very prominent and influential Christian and Messianic leaders of the 21st century. Unfortunately, some of these leaders behave or talk as if it is God who is giving them those heresies as a prophecy and/or as if the unbelievers are the only ones who misunderstood the Biblical creation story.

The enlightenment I got concerning the creation story helped me to better understand that no human being can claim that he or she knows everything, nor that everything anyone says comes from God, and no human nor even prophet has ever existed who never made a mistake! Don't be fooled by people who tell you that they know everything about you or that God talks to them all the time.

In the end, I realized that some of the things I said in this book were not supposed to be known by men in this world, or in this age or era, but after their death. However, I had some life experiences that I cannot relate here, which allowed me to tap into some supernatural realms of knowledge and understanding. Hence, I was able to craft and download certain ways of thinking, which permitted me to prove things that a mere human being cannot actually establish nor understand!

7.3 What if your rational questions against the existence of God can be proven wrong someday?

Because you are still with me, reading this book until this point, I really thank you for your effort to listen to my argument until the end before you finalize your decision. However, I know that some readers of this book may still be doubting God, even after reading this book. But what if they are asking those rational questions because they are just seeing the world from a perspective different from that from which God sees it? Now that I have shown you how analyzing the scientific data from a turbulence perspective, which I discovered, landed us exactly on the 3500-year-old story of the Biblical creation, why don't you explore more of what the God of Israel is saying?

People have lived and questioned the Bible in the past, and today they are no more. Staunch opponents of the Bible who are dead could have never imagined that a day like ours would ever come when a human being born and raised in the Republic of Benin, a small country in West Africa, could decode the turbulence at the beginning of the universe and come up with a scientific account that perfectly aligns with the Biblical story of the creation of celestial bodies.

- What if many things you may still be denying could be scientifically demonstrated some day after your death?
- What if there is a prize to get or a price to pay depending on whether you accept or reject the Biblical message, which also says that there is another life after death, and that where we will find ourselves after death on this earth will depend on some (spiritual and intellectual) choices we make today while we are still alive, and which cannot be completely demonstrated scientifically?
- What if this life we are living is not meant for us to understand some of the hard questions you may still be asking today, which may be causing you to deny the God that science is saying exists?
- What if this life is a test that you are taking, and God, who is the teacher, cannot give you the answers to all the exam questions yet, but only after the end of the test, which may be after your death?
- What if you could answer most of your burning questions by signing up and attending Science180 Academy (www.Science180Academy.com), which I founded to help people scientifically answer their deep questions about the origin of the universe, of life, of chemicals, and of other things?

As you can see since the beginning of this book, I did not try to insult you or try to hurt you, but I did my best to explore the origin of the universe from your perspective so I can better help you to see it from a perspective that you may have rejected or ignored. Even if you still deny God, I hope you now know that, at least, the probability of His existence is not zero. And because the probability of God's existence is not zero does not mean that God cannot exist, but that He can exist, and you may be at a loss someday if you continue to deny Him despite the mathematics and science proving His existence.

Not because the probability of occurrence of an event is extremely rare can you deduce that such an event would never happen. In other words, it is not because, in the view of the skeptics, the probability of the existence of God seems low or null according to their perspective that you can think confidently that God does not exist. People have postulated many things and used various methods to prove God's existence, but no one has ever used the approach I used in this book, which is an introduction to many other discoveries I made about the origin of the universe, of life, and of chemicals.

From the beginning of humanity until 2025, the year I, Nathanael-Israel Israel, published 9 groundbreaking books on the origin of the universe and its content, nobody has tested or scientifically proved God's existence like I did. In fact, most Christians just believe in the Biblical creation without even trying to scientifically demonstrate it. Those who tried to demonstrate it landed themselves in some weird pseudoscience, which they unfortunately think is science. Some top leaders even said that science can never prove God. But for smart scientists who can be impartial and courageous, it is fair to say that most Christian attempts to scientifically demonstrate Biblical creation have not been really scientific in the eyes of the freethinkers. It is not because you belong to a certain group that you must agree with everything that group is saying to defend its existence or the viewpoint of its members. Sometimes, you don't need to defend yourself or everything you believe or know. If you cannot demonstrate something, it is okay to say that you lack the proof, yet you can still continue to believe in it if you want, provided you are also aware of the eventual consequences of that belief. For the lack of a scientific demonstration of something does not equate to that thing being false. In other words, because science has not proved or will never prove all aspects of the Biblical creation does not mean that that story is false. Likewise, because Christians have not been able to scientifically demonstrate the Biblical creation until now according to the standards of the secular world does not mean or should not mean that Christians fail or that they are believing in something that is false, or that non-Christians are doing better than them, or that evolutionism and the Big Bang theory are true. And that is the mistake many are making. Science is not the standard by which all realities and truth must be measured. For there are some realities and truths that go way beyond science, which is very limited, at least by the spiritual and the physical that science has not reached yet. Not because Christians did not prove the Biblical creation according to the criteria of nonbelievers that God does not exist. No matter the amount of proof given to some people, they may still deny the facts. But my work proved for the first time in history that the scientific and mathematical information contained in the Biblical account of creation is accurate:

- During the formation of the universe fluids flew and were separated
- The Earth was completely formed on the 3rd day since the beginning of the formation of the Solar System, which is the beginning of the formation of the universe
- The Sun and the Moon were indeed formed on the 4th day
- The celestial bodies went through some processes before getting their final form; hence, the mention of the formation of the heavens and earth in the opening verse of the Bible is correct and points to the precursor of the heavens and earth we know today.

8. WHY IS SCIENCE180 TEST OF GOD'S EXISTENCE UNIQUE AND DIFFERENT FROM ALL EXISTING GOD THEORIES INCLUDING CREATIONIST THEORIES?

- How will this book make you happy and improve your life and goals?
- What will you gain or save by reading this book and applying its principles?
- How will this book help you to do things better to get closer to the best version of yourself that you deserve?
- How can this book help you to overcome, alleviate, or eliminate your God's existence pain points we discussed earlier?
- How does Science180 creationism best suit your needs, and what makes it stand out from the other theories specialized in universe origin and God's existence?

If you are interested in the answer to any of these questions, you are at the right place. Just sit and relax.

8.1. What can you get from this book regardless of your belief or doubt?

Whether you are a believer or a nonbeliever, by the time you reach this point of the reading, I hope you agree with most people that this book actually presents something that no one else has ever offered pertaining to the scientific proof of the Biblical account of creation. As you have seen in the previous chapters, based on my discoveries on the origin of the universe, I demonstrated (using facts that can be tested scientifically) that not only was the universe visible to human beings NOT formed after billions of years of processes, and no process connected to its formation ever lasted billions of years or even millions of years, but also that the Biblical account of creation is scientifically true.

While many people have talked about creationism and even evolutionism to try to explain our origin, no one (other than I, Nathanael-Israel Israel, the founder of Science180) has offered a complete, undeniable scientific proof of the universe's origin that allows you to properly understand how the scientific data perfectly fits the Biblical account of creation without forcing you to check your brain or faith outside the door, or forcing you to believe in things that cannot be factually demonstrated. In other words, while countless individuals and organizations profess to scientifically teach the origin of the universe, only Science180 delivers a comprehensive and universal approach to God's existence by digging into the massive scientific data and using unconventional tools to decode the secrets they contain for the benefit of all. Science180 also helps people understand, break through, and fix the limiting origin-related beliefs, theories, assumptions, and other barriers that block or slow down their journey to have a correct understanding of the origin of the universe based on facts. I will now give specific examples.

Whether you are a scientist or a layperson, here are selective benefits this book provides or will provide for you:

- Easy-to-use and easy-to-understand high-quality explanations and proofs of God's existence that finally satisfy your intense desire for freedom from beliefs and from scientific theories about God's existence that suffocate you and bind your mind, faith, unbelief, heart, and education

- Eliminate the struggles that you have faced for a long time when trying to decode how the universe came into existence, therefore helping you to gain the respect, admiration, power, and confidence you deserve while addressing God's existence problems

- Remove the worries of not being sure of God's existence, therefore helping you to sleep better at night

There is no struggle between science and the Biblical account of creation, but that it was the human interpretation of the scriptures and the scientific data that caused this apparent discrepancy.

- Eradicate the common mistakes people (even the most educated ones) make when it comes to deciphering God's existence, and use it to harness the power of the already-collected scientific data to challenge the status quo of existing theories on the origin of the universe, life, and chemicals

- Acquire the ability, insight, intelligence, wisdom, and extraordinary knowledge for interpreting difficult scientific data, explaining riddles, and solving the most complex scientific problems related to God's existence

- Find out for yourself whether what you have learned about creation or God is scientifically true or not

- Accurately decode creation mysteries hidden in scientific data collected over the ages

- Finally have the accurate proofs of how and why science is not against the Biblical account of creation, but confirms it so you can fearlessly push the boundaries of the human ability to properly understand what is perceived as un-understandable, mysterious, supernatural, unimaginable, impossible, and unthinkable that hold you back

- Fully comprehend that there is no struggle between science and the Biblical account of creation, but that it was the human interpretation of the scriptures and the scientific data that caused this apparent discrepancy

- Learn materials that will educate you about the amazing scientific truths hidden in the Biblical account of creation so you can stand tall as a symbol of freedom, power, creativity, and originality in your field of expertise

- Get a fresh perspective on the origin of the universe that defies the existing cosmological norms and traditions that may hold your true understanding of the universe's origin and God's existence captive

- Avoid the painful regrets haunting countless people, even dead ones, when they look back at how they lived their life not knowing the universe's origin and God accurately

- Disrupt all religious and scientific chains of repetitive nonsenses about God's existence and turn your attention toward the original discoveries of Science180 so you can better connect with ideas that lead to greater innovation and prosperity

- Challenge the cosmological status quo and embrace the real change that will disrupt all the cages or chains that were holding you but that you ignored

- Know for sure that God exists and He really created the Earth, the Moon, and the Sun as the Bible says

- Discover how modern science proves the inerrancy of the Biblical account of creation of the universe so you can put a full stop to all scientific and religious efforts threatening to undermine faith or to oppose it with science

- Open the world of science to the truth in the Biblical account of creation for the benefits of humankind and become a resourceful person for people seeking to understand the relation between science and the Biblical account of creation

However, just as the God's existence problems I addressed in chapter 2 depend on people's backgrounds, the specific things you get from this book can vary depending on whether you are a freethinker (e.g., an atheist, an evolutionist, or a rationalist), a Young Earth creationist, an Old Earth creationist, an intelligent design proponent, etc. Therefore, I will quickly summarize how the evidence of God's existence I presented in this book is uniquely tailored to you.

8.2. What gains creators or pain relievers can freethinkers (e.g. atheists, rationalists, evolutionists, agnostics) get from this book

If you are an atheist, agnostic, evolutionist, or any other kind of freethinker, this book helps you to:

- Get a unique opportunity to examine religion using science and critical thinking
- Learn about a comprehensive scientific theory of creationism that doesn't just criticize evolution or the Big Bang theory, but that uses modern science to rationally prove that God created the universe just as the Bible says
- Break free from the suffocating expectations of some theories that have hijacked your belief or doubt about the existence of God
- Learn to detect, correct, and remove all misinformation, ambiguity, and misleading claims and theories surrounding the existence of God
- Explore the proofs of the existence of God through rational arguments or empirical evidence with real scientific data that you desire
- Delight in reviewing your opinions about the Biblical account of creation based on reason and rational decision-making without feeling influenced by tradition, authority, established belief, or religious pressure
- Logically examine the existence of God, not by just agreeing that excellent things in nature prove God's existence, but by checking tangible, scientific facts you can rationalize in your mind
- Scrutinize and discover how science does not threaten or undermine faith, but confirms it
- Understand that there is a rational justification for believing in God and that religious people are not stupid or delusional, as some atheists say!
- Help you to understand the evidence of the existence of God and willingly decide to believe in Him if you want without any religious pressure
- Answer your doubts about God's existence and gain confidence in the true interpretation and trustworthiness of the Biblical account of creation
- Rationally explore how Science180's interpretation of nature and the Biblical Genesis narrative of creation shockingly led to a perfect match between science and the Biblical account of creation, a milestone that until recently was thought to be impossible

8.3. How can this book create gains and relieve God's existence pains for creationists?

If you are a creationist, this book offers you the following benefits:

- Scientifically validate the authenticity of the Genesis story of creation
- Become happy or happier by unlearning all controversies surrounding the Biblical account of creation
- Get a world-class Bible creation education to prepare yourself to effectively win stubborn scientists for God in the face of ungodly opposition
- Use the real understanding of Genesis to scientifically answer deep questions about the creation of the universe and set your heart at peace
- Learn Science180 creationism to confirm the factuality and historicity of the Genesis story, get scripturally sound and scientifically accurate resources on God's existence so you can skillfully detect and correct erroneous interpretations of the Biblical creation narrative
- Embrace the accuracy of the Biblical account of creation to champion and propagate a Biblical worldview that honors God, the Creator
- Science180 creationism improves the well-being of people through their proper understanding of God's existence
- Because I empirically proved that the Biblical account of creation can be tested, modified, or rejected by scientific means, you can now properly defend how Science180 creationism (that I spearheaded) can be included in the processes of science
- Receive the tools you need to efficiently teach the Biblical account of creation and accurately answer its critics and better defend your faith
- Learn how to rationally engage with freethinkers and keep the spark of their curiosity of God's existence and desire to know more burning bright
- Boldly answer difficult creation questions rather than scolding people for being impertinent or rude when they ask them
- Find the undeniable, scientific evidence for the Biblical account of creation that you have been seeking and start better engaging with creation questions because you are certain of the facts and how to prove them
- Build, rebuild, and strengthen your Christian faith by confidently knowing that your belief in the Biblical account of creation is accurate and trustworthy
- Rationally learn how to defend the Biblical account of creation against those claiming that it is unscientific
- Scientifically tear down all arguments raised against the Biblical account of the creation of the universe
- Properly resolve all conflicts that people try to bring between science and the Biblical account of creation
- Make the Genesis story more relevant and useful to the needs of humankind so people can better believe in God, their Creator

- Equip yourself with state-of-the art proofs of the Biblical account of creation so you can better defend and cover yourself and others against any legal attack from the deniers of Biblical creation
- Understand why the "science-only" mindset is as dangerous as the Biblical literalism mindset that discards all mainstream sciences, for both science and the Bible can be appropriately used to explain God to unbelievers

If you are a Young Earth creationist, this book will help you to:
- Properly expose, disprove, and debunk the silliness and the irrationality of all theories denying the inerrancy and literalism of the Biblical account of creation by the God of Israel
- Equip yourself and others to argue in favor of the teaching of the Science180 creationism in public schools so lives can be saved

If you are an Old Earth creationist, this book additionally will help you to:
- Liberate yourself and others from trying to present or understand the Biblical account of creation within an evolutionary framework
- Break free from the suffocating expectations of evolutionism and the Big Bang theory that have hijacked your belief in God and your secular education
- Free yourself from the expectations of Old Earth creationism that has sequestered your mind for a long time
- Enter a new area of freedom and power that crush the head of evolutionism, Big Bang theory, and all other anti-creationist theories that have held your clear understanding of the Biblical account of creation captive for so long
- Remove scientific obstacles that hinder you from accepting the literal interpretation of the Biblical account of creation

If you are an Intelligent Design proponent, this book offers you an accurate, scientific demonstration of the design in the universe by God that you have been looking for.

8.4. What really differentiates Science180 creationism or what set Nathanael-Israel Israel's proof of God's existence apart from all other God's existence theories

In the rest of this chapter, I will present indisputable facts of the demystified intersection of science and faith that smart freethinkers are honestly celebrating with rational creationists.

- Without Science180 cosmology, which I spearheaded, who could have ever imagined that science and faith could meet on a hot topic such as the origin of the universe?
- Without my discoveries, who could have ever thought that any creation story could be demystified in the 21st century?
- Could you have ever imagined that freethinkers can celebrate anything rational with creationists that they thought were irrational?
- How could that have happened without honesty and real science strategically mixed with religion?

Who said that God used evolution to create the world? Who said that the days of creation were millions of years? Who said that it is impossible to scientifically prove the existence of God? Who said it is impossible to accurately understand turbulence? Who at the National Academy of Sciences unequivocally said that "creationism has no place in any science curriculum at any level"? Who said that the teaching of Biblical creation in the US schools is unconstitutional and unscientific? Who said that creationism cannot be confirmed by using the scientific method; therefore, it must be removed from science curriculum? ... Definitively not Nathanael-Israel Israel, nor anyone else at Science180. For at Science180, we do not believe in any of that nonsense.. With Science180 creationism on your side, you are not just learning about the origin of the universe, of life, and of chemicals, but you are embarking on a journey during which you will use science and the Bible to transform your life today and forever. Before we talk about how to start our journey together, let's first recapitulate what Science180 creationism offers you:

- Undeniable scientific argument that reconciled science and the Biblical account of creation
- Easy-to-understand proof of God's existence
- State-of-the-art decoding experience of God's existence that helps you fight wasteful programs and theories
- The most accurate, reliable, safest, best explanation of God's existence ever
- God's existence rule breaker you love

8.4.1. Undeniable scientific argument that reconciled science and the Biblical account of creation

Since the beginning of humanity, only I, Nathanael-Israel Israel, and nobody else, have been fortunate to offer a scientific explanation of the formation of the universe and God's existence that perfectly matched science and the Biblical account of creation. Indeed, unlike all other past and existing creationist speakers, Nathanael-Israel Israel is the first person in the whole world to calculate the mathematical equations that scientifically demonstrated that the Earth was formed 2.82 days after the beginning of the universe, while the Moon and the Sun were formed 3.32 days and 3.69 days, respectively, after the beginning of the universe, therefore confirming the 3500-year-old Biblical account of creation, according to which the formation of the Earth was completed on the 3rd day, while that of the Moon and the Sun was completed on the 4th day of creation. Nathanael-Israel Israel is also the first person in history to scientifically demonstrate that each day in the Biblical account of creation was literally 24 hours, a milestone that accurately reconciled science and the Bible and overturned the myth according to which countless people, including believers, have thought that each day of creation was millions of years (a misunderstanding that caused many people to deny God, the Creator). Hence, Nathanael-Israel Israel is honored to be acknowledged as the first person who ushered in a new era for the proper understanding of the Biblical account of creation and its application to decode the universe and its content for the benefit of humankind.

8.4.2. Easy-to-understand God's existence proof

Most existing theories or arguments about God's existence are complicated, inaccurate, expensive, time-consuming, and labor-intensive. Whether you are a physicist, a chemist, a life scientist, a social scientist, or any other type of expert, or even a layperson, you will find the demonstration in this book very easy to use in helping you take your career, initiative, research, curiosity, knowledge, and ability to the next level. Rather than dealing with incompatible and inconvenient books on God's existence and the origin of the universe, you can get a comprehensive understanding of the most crucial origin and God's existence questions in one place: Science180 (www.Science180.com). Grounded in empirical observations, Science180's God's existence argument targets challenging God's existence problems from many angles, using strategies that accurately break down the complex origin hurdles into a language that you can easily understand. With this book, you don't have to read many books to get the needed information on God's existence here and there, but you can get a comprehensive insight right here. This convenience and accuracy are two of the things that make "Science180 Accurate Scientific Proof of God" different.

8.4.3. State-of-the-art decoding experience of God's existence that helps you fight wasteful programs and theories

Science180 delivers innovative books and services that create an extraordinary experience for scientists, laypeople, children, and anyone else who wants to get an accurate understanding of God's existence. Indeed, the proper God's existence argument can make your life experiences enjoyable instead of giving you headaches due to a false approach, outdated perspective, or unrealistic philosophy. With

"Science180 Accurate Scientific Proof of God," you can enjoy a comfortable and hassle-free experience that leads to an unprecedented understanding of how the universe was created. Without bothering you with unnecessary and irrelevant data, formulas, or theories, Science180 books and services are friendly crafted and presented with your God's existence needs in mind. They help you to clearly use a new method involving various disciplines to decode how the content of the universe came into existence. By helping you to quickly connect or put together the complex dots concerning the formation of the universe, you will easily have a world-class experience of God's existence. Instead of having to replace outdated universe-origin books and services with new ones very often (because they were incorrect), you will take pleasure in the innovative, well-thought, long-lasting, and time-tested books and services of Science180. You can learn more at www.Science180.com/books.

Science180 creationism is a God's existence package sustainably crafted, inexpensive, and designed to fit your needs. Planning all your projects can be overwhelming and resource-consuming, especially if you are trying to make sure no resource (including money) goes to waste. Science180 creationism focuses on critical universe-origin variables that most scientists have mistakenly discarded due to their misunderstanding of the real proofs of God's existence. Therefore, unlike all other cosmological theories, Science180 creationism will help you avoid wasting your resources and time on unprofitable origin projects so that you conserve more cash and spend it on more valuable and profitable things.

8.4.4. The most accurate, reliable, safest, best explanation of God's existence ever

Most theories about God's existence don't address the critical universe-origin questions. Therefore, most existing theories are not comprehensive, but they just partially fit a limited amount of data. That is why many of God's existence theories have been abandoned and replaced by new ones, which will be replaced shortly as well. In contrast to existing God's existence theories, Science180's God's existence argument considered a very broad matrix of data and used a clear-cut approach that made its explanations reliable and applicable for all disciplines. Science180 helps you to safely embrace its authoritative findings, knowing that they are holistically correct. Science180 has everything you would want to know in a premium and valid argument of God's existence. The complex data it examined and its unique perspective will help you to enjoy properly decoding, playing with, and mastering God's existence like never before. With an ultra-powerful ability to explain difficult concepts related to the formation of the universe, using simple terms and an unmatched accuracy, Science180 creationism helps you cut useless costs and time-consuming research related to your efforts to unambiguously understand the existence of God, while helping you to stop wasting resources on useless and unrealistic theories that will yield nothing. By doing so, you can focus your efforts on real money-making problems that can make your initiatives more profitable, taking yourself and your organization to a higher level, where new groundbreaking doors will open for you, while you outpace your competitors who are still struggling to accurately understand the beginning of the universe and prepare for a better future.

8.4.5. God's existence rule breaker you (will) love

Science180 doesn't play by the ancient or modern rules set by some people who have tried to explain the origin of the universe using theories that refuse to acknowledge the real meaning of the scientific data. Instead, under the leadership of its founder, Nathanael-Israel Israel, Science180 used the data collected throughout the ages to scientifically decode the origin of the universe from a perspective that will surely help you, free you, empower you, challenge you, motivate you, and cause you to dance to a fresh and original beat that takes you many steps closer to your best life today and forever.

- At Science180, are we scared of what the established orders will say? No.
- Do we fear being controversial and original? No.
- Are we concerned about the useless theories we will dethrone or destroy? No.
- Are we concerned about sticking out our tongue at traditional and secular universe-origin, life-origin, and God-inexistence nonsense? No.
- Are we concerned that some people may not like us? No.

We just focus on our message and our mission—properly decoding and sharing the proof of God's existence—knowing that by doing so, many people will get the needed priceless help.

> The problem people have at the interface of science and faith is connected to how they interpret the scientific evidence using religious materials or doubt.

People are tired of the overwhelming number of books written on God that are not really addressing the critical universe-origin problems. Some of this fatigue is because people don't know which books are correct or which ones will satisfy their God's existence needs. By customizing this book (and other Science180 books and services) according to your needs, Science180 allows you to focus on what fits your scientific or lay background, religious or nonreligious views, and belief or disbelief regardless of your age. Below, I addressed six selected six-figure mistakes that most people, including Christians, make and how to avoid them using Science180 creationism breakthroughs:

- Unlike most creationists, who usually start with religious texts to make their case for the existence of God or to show how science can explain their belief, and unlike the proponents of Intelligent Design, who usually start with natural empirical results to draw intelligent inferences from the evidence—

not to prove God, but an intelligent designer—Science180 creationism is the first and only argument for God's existence that started the demonstration of the origin of the universe with scientific data collected by renowned secular scientists and then landed on the scientific proof of the creation of the universe by God as summarized in the Biblical account of creation.

- With Science180 creationism, you don't have to sacrifice or throw away the scientific and rational quality of your mind to receive God or believe in the Bible. Science180 creationism proved beyond a doubt that you can know God and believe in His Word by using your scientific and rational mind 100%. I have discovered that the problem people have at the interface of science and faith is connected to how they interpret the scientific evidence using religious materials or doubt. Through Science180, I help people of all backgrounds to properly navigate the challenging issues related to the universe's origin so they gladly live nonstop.

- Science180 creationism rejects the allegorical interpretation of the Genesis account of creation and that the days of creation cannot be viewed allegorically.

- Unlike what some people think, Science180 creationism proved that each creation day doesn't symbolically represent 1000 years of the world's history

- Unlike day-age creationism, Science180 creationism proved that each day of creation was not longer than 24 hours.

- Contrasting gap creationism, Science180 creationism proved that there is no gap of time between the days of creation or between the first verse and the second verse of the Genesis story in the Bible.

- Opposing Progressive creationism, Science180 creationism showed that God did not create the universe gradually over hundreds of millions of years

> Because Science180 creationism explanations can be tested, modified, or rejected by scientific means, they can be included in the processes of science.

- Unlike Old Earth creationism, Science180 creationism proved that the Biblical account of creation is a strict chronological record of how God created the universe and its content.

- Unlike Old Earth creationism, Science180 creationism exposed that the Genesis story is not a topical order of events.

- Opposing day-age creationism, Science180 creationism scientifically demonstrated that the days of creation were ordinary 24-hour days that do not represent "*much longer periods of millions or billions of* years."

- Unlike what Gap creationism denies, Science180 creationism has proven that Genesis 1:1 is part of the first day of creation and is not part of an "*infinite antiquity*" that ended when the Spirit of God moved upon the face of the waters.

- Because Science180 creationism explanations can be tested, modified, or rejected by scientific means, they can be included in the processes of science.
- Unlike Intelligent Design, Science180 creationism does not point at the creator of the universe as a simple designer or engineer, but as the God of Israel, the God of the Bible who spoke to Moses, the major prophet who wrote the Biblical account of creation in the Bible's book of Genesis about 3500 years ago.
- Science180 creationism proved that, just as the first verse of Genesis said, God created the Heaven and Earth in the beginning, yet it took some time before the Earth could get its final form (which was completed on the 3rd day), so also at the end of the 6 days of creation, everything in the universe was created, but it took a certain additional time for some of them to be fully formed and get their final shape. For instance, Science180 creationism, which proved that the Earth was formed on the 3rd day, while the Moon and the Sun were formed on the 4th day, also proved that Jupiter was not formed yet at the end of the 6th day, but its precursor had to go through some processes (which took a few more days) before Jupiter as we know it today was formed about Day 14 (details for the birthdate of celestial bodies can be seen in Nathanael-Israel Israel's book called *"Turbulent Origin of the Universe"*).

- Unlike Young Earth creationism, which has a better perspective than all existing creationism theories before the days of Nathanael-Israel Israel, but which never actually demonstrated the Biblical creation scientifically, Science180 creationism (spearheaded by Nathanael-Israel Israel) actually properly demonstrated the Biblical creation to a scientific standard that considered the real needs of freethinkers, therefore making countless freethinkers fall in love with the God of Israel they have denied.

> Science180 creationism exploded the myth of those who think that the days of creation in the Bible were millions of years instead of 24-hour days each.

- In short, for the first time in history, Science180 creationism exploded the myth of those who think that the days of creation in the Bible were millions of years instead of 24-hour days each. It also scientifically challenges and refutes the belief of the Big Bang theory and all other theories that think that there is no God who created the universe. It proved that the God of Israel is the one who created the universe. It also suggests that this proof is not an endorsement of everything Israel has been doing, for except for the Messianic Jews, most of the Israelites have rejected God, even until today. Yet, the nation of Israel is the apple of God's eye, the timetable of the world that informed people should look at to know where we are in the history of the world: the end is near. I addressed that topic in other incoming books (learn more at www.Israel120.com). Science180 creationism also highlights the message of the grace of God, without which no one could be saved, not even the popular preachers or prophets who, just like some unbelievers, have made gigantic mistakes about creation to the point of embracing anti-creationist theories while holding the Bible in their hands.

- Science180's God's existence proof scientifically defies all existing theories that rob God of his glory of creating the universe and its content. It let you harness proven universe-origin facts to holistically pave your way to higher ground and properly reject all pressures to scientifically or philosophically fit in with anti-creationist models. It educates you to better resonate with people who are craving something original that breaks wrong explanations of God's existence and the origin of the universe. Science180 creationism will finally satisfy your desire for freedom from beliefs and scientific theories about the universe's origin that suffocate and bind your mind, faith, unbelief, heart, and education. It can help you to harness the power of the already-collected scientific data to challenge the negative status quo of existing theories on the origin of the universe and its content. It will help you and others to break free from the suffocating expectations of evolutionism and the Big Bang theory that have hijacked secular education for a long time. It will also help you and others to break free from the dangerous evolutionist ideas that some forms of creationism have infiltrated into the minds of some believers for a long time. It can help you to light an unquenchable fire under all traditional and scientific nonsense about the existence of God and the creation of the cosmos. In short, it turned all existing cosmological theories and all incorrect established norms and absurdities of the universe's origin on their head by accurately reconciling science and the Biblical account of creation for the first time in history.
- Science180 creationism teaches people how to scientifically unlock, believe in, and interpret the Biblical account of creation without checking their scientific or rational brain at the door. It spreads very accurate, rare, and factual truths (about the origin of the universe, life, and chemicals) that will save you time and money and improve your life today and forever.
- Science180: When science meets the Bible, a major door is open to win souls for Christ instead of hating them.

So far, I talked a lot about my book *"Turbulent Origin of the Universe,"* but in fact, I wrote another amazing book called *"From Science to Bible's Conclusions,"* tailored to laypeople, in which I laid out my discovery in a format that all nonscientists and scientists will enjoy. *"From Science to Bible's Conclusions"* is what I called the "public version" of my book on the origin of the universe, and it is custom-made for the general public. It is a great summary of the scientific version from a perspective that laypeople will fully understand. In this book you are reading now, I told you the story about God's existence, but "*From Science to Bible's Conclusions*" is a quicker summary of the long scientific demonstration I did in "*Turbulent Origin of the Universe*" in a language that laypeople will easily understand. In "*From Science to Bible's Conclusions*," I broke down the complicated (scientific and philosophical, including religious) data about the origin of the universe in a simple language that the general public can fully understand and know in order to live happily forever. Quickly grab and read this scientifically verifiable, bestselling book to finally get the accurate, jaw-dropping answer that has been rationally shaking believers, skeptics, and all freethinkers. Don't wait! Learn more at Science180.com/public. Therefore, if you are not a scientist who can quickly read and understand 700 pages of a scientific document, I highly recommend that you get "*From Science to Bible's Conclusions*" first. Even if you are a scientist, getting that book will also help you to have a quick scientific glance at my discovery on the origin of the universe. Remember that in this book you are reading now, I just focused on the existence of God, which is a small portion of my discovery of the origin of the universe, of chemicals, and of life.

Knowing the value of this book, I realized that it would be unfair not to write a children's version of it. But how can I present my discoveries to children in a language they will admire? As you may know, the most difficult part of writing scientific things for children is how to break down complex technical concepts into simple words that they, and even anyone who can read, can clearly understand (without losing the accurate details and facts). When the topic to address is about the origin of the universe, the task is even more challenging for most people, but not for me. I have spent years breaking down my discoveries into a book called "*How God Created Baby Universe*," so children could enjoy it as they also answer most of the children's questions related to God and the Biblical account of creation. As long as a child can read, he or she will find that amazing book extremely helpful to grasp all the complicated concepts needed to properly crack the origin of the universe in a language that all children ages 7-12 and anyone who did not go very far in school can fully comprehend.

I also wrote another book called "*Origin of the Spiritual World*." That book is what I call the pseudepigraphic, or the hidden version of my books on the universe's origin, and it is meant for believers who want to tap into a higher level of scriptural secrets that most people may not believe. That book draws the attention of the world toward the pseudepigrapha (a collection of hidden and rejected books, yet filled with deep secrets still valuable today) and explains how, for thousands of years, God has already revealed deep details about the supernatural origin of the universe, but people (including those who believe or claim to believe in Him) have just refused to literally accept God's mysterious story of creation, which can never be understood by just sticking with conventional science. If you believe in God, have some origin-related questions whose answers you cannot find anywhere, not even in the Bible, and want to tap into historically neglected revelations to answer fundamental universe and life questions, then be sure to get a copy of "*Origin of the Spiritual World*" today. Learn more at Science180.com/pseudepigrapha

In short, I, Nathanael-Israel Israel, broke down my discovery about the formation of the universe into many books so that you, the readers, can pick the ones that correspond to your needs and interests without disappointing you or wasting your precious time. These books come in many versions (e.g., scientific version, public version, chemical version, biological version, biblical or prophetic version, pseudepigraphic version, and children's versions), targeting people according to their expertise, educational background, and interests. If you want to have the entire big picture of my discovery of the origin of the universe, life, and chemicals, and to enlighten your life and career, then plan to get all or some of these books that best suit your needs and interests. For more details, visit Science180.com/books.

'Science180 Academy' Success Strategy
SCIENCE180 SEMINARS

People whose awareness is raised by Science180 usually ask me to go deeper or they wonder "what's else?". That is one of the reasons Science180 trains them through strategic work sessions (during seminars or training sessions) that transfer customizable skills and solutions to them. Science180 Seminars are client-centered and tailored to strongly engage the clients so they maximize the discovery of and the tapping into new opportunities, and exponentially outperform their expectations. Science180 offers customizable seminars that can be labeled as a colloquy, conference, consultation, discussion, forum, keynote speech, lecture, lesson, meeting, symposium, summit, study group, tutorial, workshop or working section accordingly on any topic related to:

- Universe-origin for scientists and mathematicians, philosophers, laypeople, and the general public
- Universe-origin or universe creation for believers
- Life-origin for life scientists, for all other scientists, and for believers
- Chemical-origin for scientists
- Universe-origin seminars for children
- Universe and life-origin for pseudepigraphic believers

As you contact us with your needs, we can customize your program accordingly. Learn more at Science180Seminars.com.

'Science180 Academy' Success Strategy
SCIENCE180 CONSULTING

Because Science180's trainings, seminars, or strategic work sessions (through which it transfers skills and training solutions) are great, some customers want to go even deeper on a long-term, sustainable basis. That is where Science180 Consulting, one-on-one consulting, and mentoring (that some people may prefer calling coaching programs) comes in. That is where Science180 can truly change people's behavior on a long-term basis according to their specific needs. With Science180 Consulting, you will discover and understand the deep secrets of the formation of the universe, life, and chemicals around you. Hear Dr. Nathanael-Israel Israel's personal selection and teaching on key topics that will help you break the code of the universe formation and functioning. All strategically designed to enlighten you, guide you to navigate and filter the massive data collected on the universe and its content so you know how to answer the world's most challenging origin questions, remove any scientific and philosophical cataracts that may be blocking you, and help bring you many steps closer to your best life today and forever. Science180 Consulting will train you, transfer unconventional skills to you and change your behavior so you go deeper. To get started today or to learn more, go to Science180Consulting.com.

'Science180 Academy' Success Strategy
SCIENCE180 MASTER CLASS

Hear the greatest scientific and philosophic lessons from top scientists, philosophers, thinkers, and public figures who have realized historic mistakes they made in life (concerning the origin of the universe, life, and chemicals), and that they corrected thanks to the historic discovery of Nathanael-Israel Israel, who founded Science180 and who is known as the one who truly decrypted the universe origin for the first time. In their own words, these renowned personalities share with the world key lessons they have learned in life and how people can learn from their experiences to improve lives instead of repeating their mistakes that many people still ignore at their own perils. To learn more, contact us at Science180.com/contact.

9. WHY AM I TEACHING THE WHOLE WORLD HOW SCIENCE ACCURATELY SUPPORTS FAITH AND A BIZARRE CREATION STORY … AND WHAT CAN YOU DO TO AVOID THE DANGEROUS THING THAT WILL HAPPEN NEXT?

Top creationists and anti-creationists were asked to explain in 1 word what will happen in the world next. They all gave the same shocking answer. To know what it is, read this chapter.

Indeed, after reading this book, I know you will still be left with many questions, some of which I have already addressed elsewhere, and others that I may never know but against which we can work together. Some of them will be tackled by other people, and others may never get solved. Some I have thought about them, but could not find an answer or could not share my thoughts on them yet. Below, I will share the secrets behind some of those questions with you.

9.1. Twenty four questions you cannot ignore if you want to decode the leaked secret about what will happen next in the world

Believe it or not, the world is not going to stay the same after the publication of this book and 8 others I released in 2025. Because you are my friend and I want to serve you, before I wrap this book up, I would like for you to ponder on what may happen next, which is encrypted in the following 24 questions you can set 24 minutes on:

1. Will Christians use Science180 creationism to take evolution and the Big Bang theory to the court and present God to smart atheists and freethinkers in a way they will quickly like?
2. Will evolution be banned from public schools, or can the introduction of God in public schools abolish or impede science?

3. Will Young Earth creationists finally understand that what they were calling science is a kind of pseudoscience that they need to abandon to embrace Science180 creationism so freethinkers can better listen to them and enjoy God?

4. Will Old Earth creationists abandon their evolutionary love story, or will they persist in their error and use Science180 creationism to deny the God that many atheists have started loving after reading this book?

> If the suffering of some people today is not sufficient to allege that their ancestors or great grandparents never existed, can you be right to use the suffering of people in the world to claim that God, the creator, never existed.

5. If the suffering of some people today is not sufficient to allege that their ancestors or great-grandparents never existed, can you be right to use the suffering of people in the world to claim that God, the creator, never existed, despite what I, Nathanael-Israel Israel, demonstrated in this book and others I published in 2025?

6. Will creationists apologize to freethinkers so the latter can finally rationally listen to the message of salvation in Jesus, or will Christian leaders, including political leaders, use the discovery of Science180 creationism to crush the unbelievers who need to know God willingly?

7. Will Christians use my findings to justify their vengeance toward the secularists who have cast God, the Creator, out of the public systems for a long time? If God is happy with my discoveries, can He be really mad at everything in secular science, which collected the raw data I used to prove His existence?

8. Will any world leader take an executive order to force the teaching of Science180 creationism in public schools to "honor" God and make the Bible famous? I wish to use the term "Make God (Jesus Christ) Great Again," but I can't, for God never lost His glory or greatness, and He is greater than what we may think or imagine as great, and we cannot make Him great, but we can do something better, nicer, and lovely to help the lost and the needy to know Him, and even more, rather than hating them or chasing them away from the real discussions targeting the needs they ignore.

9. Must Christians apologize to atheists, rationalists, and all other freethinkers for the wrong proofs that creationist scholars and preachers have used to argue biblical creation scientifically or pseudo scientifically?

10. Because the Biblical creation has a scientific power that can help reorient science, are you frustrated by those who think that Science180 creationism must be taught in public schools for the benefits of all?

11. Can anti-creationists expect creationists to ignore Science180 proofs of God's existence and just continue to let evolutionism dominates, dictates,

and rules the curriculum in public schools, while God the creator is cast out?

12. With all the proofs of God's existence I provided in this book, do you think it will be fair to use Science180 creationism to reopen science classrooms to discussions of God, the creator?

13. Will anti-creationist scientists take seriously the discovery of the perfect match between science and the Biblical account of creation?

14. Will Christians sincerely embrace Science180 creationism and use it to legally "fight" for its teaching in science classes, or will they just continue to wish that another form of creationism will reopen science classrooms to discussions of God someday?

15. Will school boards, policy makers, and administrators finally allow the teaching of Science180 creationism in public schools and help embody it properly in doctrines and regulations, even political laws, so God can bless our land, or will they reject Him until He judges the land?

16. Will Young Earth creationists continue to just believe that the Earth is young but refuse to embrace Science180 creationism, which actually proved for the first time in history how God created the universe?

17. Can we deny that the universe or the Earth is billions of years old but refuse to embrace the math spearheaded by a Beninese-born American scientist that says that God created the universe as the Bible says?

18. Will Old Earth creationists finally reject that God has used evolution to form the universe and accept the literalism of the Biblical account of creation, including that the days in the Biblical book of Genesis were not geologic ages but were six twenty-four-hour consecutive days of creation?

19. Will Old Earth creationists stop interpreting the Genesis account of creation allegorically so they stop embracing evolutionism in the name of the Biblical creation misunderstanding?

20. Will Intelligent Design proponents finally understand and publicly acknowledge that the intelligent designer in their theory (which they don't publicly call God) is actually the God of Israel and know that science has identified that the intelligent cause of the formation of the universe is supernatural?

21. Will Intelligent Design proponents publicly embrace and join Science180 creationism to name the intelligent designer of the universe as the Judeo-Christian God, even if their model won't be accepted by the secular community or even if others think that they have reduced the Creator God to a mere engineer?

22. Can Intelligent Design proponents continue to refuse to publicly disclose the intelligent designer as the God of Israel revealed in the Bible and think that their strategy can increase the acceptance of the intelligent design theory, while Science180 creationism has already corrected, debunked, or demystified all existing creationist theories?

23. Can we continue to ignore the intersection of science and the Bible while still

expecting to build the holistic support needed to advance humankind without letting modern science lead us to a place we won't regret?

24. Did or will some Christians complicate the rational pathway that some atheists and other freethinkers could use to scientifically understand God and believe in Him, or will atheists persist in their direction and brace for the impact they could not see yet?

I would like to hear your thoughts on these questions. Please visit Israel120.com/feedback to give me your feedback. I cannot provide a comprehensive answer to all of those questions in this book, but, in the next segment, I will boldly share with you what I can for now.

9.2. My unprecedented warnings to the whole world that you cannot afford to ignore if you want to be happy and avoid the deadly mistakes most people make

After sacrificing 12 years to write this book and many others on the origin of the universe, of life, and of chemicals, I cannot conclude it without finally answering the most important question I know you have in your heart since you read my demonstration of the perfect match between science and the Biblical account of creation.

1. Can the world and the debate about science and religion ever be the same again?
2. What will happen next after this discovery? A revival? A revolution? A revolt? A mass conversion? Persecution? A religious war?
3. Will they maybe give me a Nobel Prize to honor God, or will they try to kill this "Black genius"?
4. If some unbelievers and even some creationists that I torched in this groundbreaking book will try to destroy me because I rocked their boat, does that scare me?
5. Do I care when I already know that some people will love my findings and others will hate them,
6. Do I believe that Science180 creationism will be taught in public schools very soon? Is this what some powerful world's leaders were waiting for to bring the Bible back to public schools, or to change the debate about the separation of the state and church, or to brush aside the separation of church and state, or to reset the religious liberty?
7. Now that we scientifically know that God exists, can anybody logically predict what some political leaders will do with that in respect to casting out anti-biblical teachings, witch hunts, and hoaxes from public schools, and

bringing the Bible and public prayer back, even when most believers don't properly pray in their prayer closet (which is nonexistent in most cases) and regularly fast in secret more than they want to publicly display?

8. Will this be a political stunt or the beginning of an authentic revival that will really save lives, even if neither our economy improves as wished nor our political party gains anything?

9. Is it maybe time to Make the Bible Great Again in Public Schools? But is it what God is just looking for, while some of those who vehemently advance those policies are privately betraying God in their heart and bedroom more than they publicly proclaim with their physical mouth to please their audience or market?

10. Can any leader be able and ready to do that in our modern world without seeking a personal gain or retaliating, even without avenging God, who said vengeance is his and who has been silenced for centuries in most public education systems? Yet He stayed quiet—or was He speaking loudly but we failed to hear Him, or are we privileged to be living in such a historic moment like this that some people are taking for granted?

11. Can this really be done fairly and legally without angering God, who also cares for the future of the nonbelievers that He loves but that some believers tend to hate because these nonbelievers annoy them a lot—or don't they?

12. Did I sacrifice my scientific career in the USA, where I graduated first in my class of hundreds of doctorates, to devote the last 12 years (2013-2025) to decoding the origin of the universe, of life, and of chemicals and to arguing a scientific case for the existence of God in vain? Why did I stay quiet and hide my findings from the public and wait until 2025 before I launched the 9 books I wrote on these subjects? Can you guess my next move? To be in the know, register for my newsletter at www.Israel120.com/newsletter.

My friend, the stake is higher than those things I mentioned above, and if you want to know, then don't put this book down until you finish at least this chapter. Now, listen to me. What I am about to say now is extremely important, and you don't want to miss even a single line if you really want to know the secret behind what is coming up next that THEY don't want you to know. Who are they, you may ask? Let's figure it out now as you seriously ponder on the 18 historic reasons that I have specifically for you:

1. I know there is a way for atheists, rationalists, secularists, and freethinkers to scientifically survive and heal from doubting the God of Israel. For instance, the dislike of the Biblical account of creation is one of the leading causes of the doubt that atheists, rationalists, secularists, and all freethinkers cast on the existence of God. The inability of most believers to support or tolerate this doubt while scientifically answering their critics, who think nothing rational can come out of the Bible, has broken the relationship between

believers and rational nonbelievers—or maybe that's how it should be? As for me, there is a proven way to engage creationists and non-creationists in a productive conversation that can progressively help atheists, rationalists, secularists, and freethinkers to view the Biblical account of creation through a scientific lens that can remove some of their doubts about God, which most believers have failed to scientifically demonstrate. If I have not convinced you yet, interview me to learn more about how creationists and anti-creationists can survive the pains they caused to one another in a way that honors the truth and improves lives. If you want to work with me to get there, I am here to help you. Just contact me.

2. If you missed the scientific formula introduced in this book, you could get the universe's origin and God's existence wrong. Why is that? Indeed, the literature is filled with scientific and philosophical theories attempting to explain the origin of the cosmos. Some people trust Albert Einstein and his relativity theory, others still stick to Isaac Newton and his gravitation theory, and others are trying to trailblaze new scientific approaches. For others, the answer is to just believe in religious books, while others believe in everything or in nothing. Nevertheless, even for those who embrace any of these alternatives, there are still challenging questions that are unnoticed or neglected at the peril of always scientifically misunderstanding the prevailing factors at the beginning of the universe. For instance, having written several scientific papers and published in high-impact journals, I know very well how to conduct and report scientific research. I could have written this book following such a conventional pattern, but to keep up with the truth of how I got here, I did not follow the traditional methodology of scientific research. I can tell you what to do to scientifically sift through the gigantic amount of cosmological and religious data to rationally crack the universe's origin and how to explain that to your team, entire family, including children, in simple and accurate language they can understand.

3. If we are not careful, the "war" between mathematics and physics that most people ignore will drift science and humankind in the wrong direction forever. Indeed, scientists have collected a lot of data on nature based on specific hypotheses, but because these data did not yield the answers they were looking for, they have been parked, and newer scientists rarely revisit them. Mathematicians, who don't need to collect new data in the field before pondering on what was already collected, take some of those old data and think about them without any realistic framework, but just by considering the path that their mind leads them through no matter how far their demonstration can lead them and no matter how real or imaginary their

findings would be. Once they come to their conclusions, mathematicians sometimes find imaginary formulas, which, when submitted to some mathematic operators, yield products, or results, that appear real, yet they are fake. Unaware of the imaginary aspects of those mathematical formulas, most scientists embrace the pseudo-real or pseudo-imaginary mathematical explanations of the real physical world as real. Yet, mathematicians don't care about empirically testing these formulas in the real world; for as long as these formulas are logically constructed, the mathematicians are ok with them. Incapable of formulating most of those mathematical equations by themselves, most scientists just embrace these formulas as total reality and start testing them by collecting data on them using some of the same mathematical framework used to build them. In the end, scientists are trapped in a circle that bonds them to never get out. And this needs to stop now! Interview me or invite me to your organization to explain this mystery to you. You can also learn more by joining Science180 Academy (www.Science180Academy.com), where I will answer your questions on the matter.

4. I am here to tell the whole world that you don't have to embrace evolution or deny God to scientifically prove that God created the universe in 6 literal days. Indeed, most Christians and atheists think that it is impossible to ever demonstrate how God created the universe. Some Christians compromisingly embrace evolutionism and other theories that oppose the literalism of the Genesis account of creation. Some Christians think that the days of creation were not 24 hours each but a period of a million years. Some Christians embrace Intelligent Design that "refuses" to publicly acknowledge God as the creator, whom they point to as the designer. In the midst of all that, freethinkers and rationalists completely reject any notion of the existence of God because, for them, science and religion cannot be reconciled to prove that anyone quickly created the universe or life in a matter of days. Unable to scientifically prove the Biblical account of the creation of the universe, most Christians think that they need to reject the literal meaning of Genesis in order to explain the origin of the universe. From experience, I, Nathanael-Israel Israel, know and have proved that none of these myths is true, and, using science, I will demonstrate it to you and your audience in a language that will fit your needs. I will scientifically explain to you "Why you don't have to embrace evolution or deny God before scientifically proving that God created the universe in 6 literal days" or "Why you don't have to reject the literalism of Genesis to convince people about creation." Although most educated people I serve, including most of those who attend Science180 Academy (Science180Academy.com), believe they need to hide their faith in order to think like a scientist, I know from experience that this is false. I will teach you "Why you don't have to scientifically check out your brain before

rationally proving that God created the universe." Most Christians think that they must use/start with the Bible in order to prove to rationalists and freethinkers that God created the universe. From experience, I know that this approach can be weak when you deal with staunch atheists and freethinkers whose minds are fixated on science and rationalism. I will show you "Why you don't have to always open the Bible before scientifically demonstrating that God created the universe." Invite me on your show or to your organization to figure it out.

5. Countless Christians think that if they just believe in God, go to church regularly, read the Bible very often, and reject mainstream science, they will someday be able to scientifically explain the creation of the universe by God to all freethinkers. As for me, that is not true. Christians need to first learn how to use the raw scientific data hidden in wrong scientific theories to create the ultimate scientific sauce or theory that accurately explains creation and its author, God.

> Christians must embrace the rational criticisms as opportunities to share the real truth with their attackers lovely instead of just attacking back the evolutionists or secularists who are criticizing them using arguments that we now know are false.

6. To some extent, some Christians are refusing to believe in the God they want the nonbelievers to accept. You may ask me why. For instance, many Christians reject the foundation of the Bible (creation), but they want the rationalists who can easily think through things scientifically to accept their creationist nonsense. Some preachers lie and even blaspheme while talking about creation, yet they think that freethinkers are the only blasphemers! Likewise, increasing numbers of televangelists and renowned ministers are preaching nonsense about God and creation. If not, how can a preacher deny God the glory of creation and attribute the formation of the universe to evolution? I am here to put the whole world on notice about why this attitude is extremely dangerous for the spreading of the Gospel and share with you what can be done to fix it. You may not like that I put it this way, but I am here to tell you that this rubbish needs to stop now so atheists and freethinkers can enjoy the God they hate. Therefore, instead of attacking and trying to destroy their opponents, Christians must embrace the rational criticisms as opportunities to share the real truth with their attackers lovingly instead of just attacking back the evolutionists or secularists who are criticizing them using arguments that we now know are false.

7.   Just as most public schools' teachings about the universe's origin are making people doubt God, many churches, pastors, or preachers are making some unbelievers doubt God more than they can imagine. Likewise, some creationist explanations of the Biblical account of creation are making people doubt science. Some believers are fighting for the return of prayer in public places, yet they are not praying properly in the closest of their own bedroom. This garbage needs to stop now. I want to help you discover and fix creation misunderstanding threats so you can learn top secrets to eliminate dangerous and confusing problems in your life and protect yourself and dear ones. I, Nathanael-Israel Israel, want you to get out of the pile of creationist and anti-creationist nonsense and meet the God of creation face to face now before it becomes too late. To say that anti-creationists are the only ones who missed the real meaning of the Biblical creation is wrong, for even staunch creationists have denied critical things about God, particularly the literalism and inerrancy of how He created everything, just as the Bible says. Therefore, no matter the source of all creationist nonsense, they need to stop now. Contact me now with suggestions of how we may work together to solve these problems.

8.   Although science has tremendously helped humankind, I have discovered the remarkable Biblical power of modern science to definitely solve the biggest interrogation mark at the intersection of reason and faith in our contemporary world in which science and religion have been positioned against each other. I can share with you how science and the Bible can work together to unlock the answer to the most-asked question of all time. I know that the word of God in the Bible can help debunk and reorient science if we can pay attention.

9.   I proved it, and I want to better help you to know that you can really be scientifically 100% sure and prove that God created the universe. Indeed, most people struggle with doubts about the scientific explanation of the universe's origin. What if you could let go of the weight of these doubts and enter a new area of freedom and power by crushing the head of and breaking free from the suffocating effects of these doubts and discover very accurate, rare, and factual truths about the universe's origin that will save you time and money and

10. I feel like some Christians need to apologize to atheists, rationalists, and all freethinkers for some stupid proofs some creationist scholars and preachers have used to demonstrate creation. Indeed, although some unbelievers willingly pushed God away, some Christians have also complicated (maybe unconsciously) the rational pathway that some atheists and other freethinkers

could have used to scientifically understand God. This trend is extremely dangerous for peace and knowledge acquisition in the world. But it does not have to be that way. Countless Christians have been denying the God they are trying to convince the nonbelievers to embrace, and to know my thoughts on what can be done to fix this dangerous trend, please contact me at www.Israel120.com/speaking.

11. I want to warn Christians and their leaders that they need to start taking the Biblical account of creation seriously, or else they will pay a huge price. Why? Because every Christian claims to believe in God, but the problem is that most Christians don't want to literally believe in the Bible, particularly in the Biblical account of creation. Why? Sad but true, it is because people's minds and modern science suggest to them that the Bible is inaccurate as they try to rationally compare what God said with what their wrong interpretation of physical data is telling them. This warning also applies to non-believers.

12. If all existing creationist theories are not corrected so their advocates can bounce back from their huge cosmological errors, some people will pay a huge price. Indeed, when people make big mistakes, dealing with the discomfort, shame, grief, and all other related pains is a component of the correction and recovery process. Sadly, most of the scientific arguments that some creationists have used to defend themselves when rationalists and freethinkers attacked them did not address how to properly correct their theory and recover from all attacks against it. Can Christians prove the origin of the universe by using the theory of relativity of Albert Einstein, which points to a direction opposing the Biblical account of creation? Come on, you educated creationists! You cannot have it both ways! You cannot use some secular theories to prove your Old-Earth creationism, Intelligent Design, and all other flawed theories, all of which are full of 6-figure mistakes that you ignore, and then reject their conclusions that deny God! I can assist you to deal with and correct the huge mistakes in the existing universe-origin and God's existence theories so people can unregretfully enjoy life. Visit Science180Academy.com today to learn more about how you can get some help.

13. By enlightening people, we can work together to stop the dangerous trend according to which many people don't trust science to lead them to any truth. Indeed, more Americans (and people of other nations as well) are increasingly losing trust in the ability of science to lead them to any sustainable truth, and this trend is particularly alarming among faith-based people, including Christians. Increasing numbers of teenagers are also dropping their faith once

they leave the traditional "covering" of their parents to head to middle school, high school, college, or university—a movement that can expose them to risking the faith of their fathers (if any) for the profit of secular knowledge that may prevent them from ever understanding the real origin of the universe or even knowing that the science they are embracing can also lead to the same conclusion as the Bible that they think is obsolete: God really created the universe in a matter of days, not through a billions-of-years process, and the formation of the Earth was really finalized on the 3rd day, while that of the Moon and the Sun was also truly on the 4th day of creation. We can work together to end this dangerous trend, which is wrongly opposing science and the Bible, or reason and faith.

14. Without angering Christians or God, there is one simple scientific formula to accurately and quickly overcome the biggest lies countless preachers have been spewing on the pulpit about creation. Indeed, although the Bible prohibited lies, and pastors preach the same, many of them have been mistakenly spreading lies about the Genesis story, maybe, let's suppose, unintentionally. Unfortunately, most churchgoers ignore those lies and take them as true, even to the point of condemning nonbelievers who oppose this nonsense just like them. Using a simple scientific formula I discovered and published after spending 12 years working on creation myths and the scientific evidence of the universe's origin, I, Nathanael-Israel Israel, will show your audience how to quickly spot universe's origin lies in any teaching and how to genuinely fix them with the Biblical truth that science accurately confirmed. Visit Science180.com/Interview to see how you can interview me to help your audience get the answers and the up-to-date information they were looking for.

15. Scarcely a day goes by that a nonbeliever does not attack the Bible and its proofs of God creating the universe in 6 literal days, yet science proved that the Biblical account of creation is correct. I am pained by the countless lives lost and the numerous people who are being misguided by the scientific and religious lies and errors. I honestly know that if I had had the opportunity to teach these people (about the origin of the universe, of life, and of chemicals, and the scientific proof of God's existence), the consequence would have been life, joy, fulfillment, and success instead of catastrophe. I showed in this book gigantic mistakes in our societies, and even those preached in the pulpit or taught in public schools, that have been drifting many away. There is a better way. I will teach you what is the one thing lacking to solve this extremely dangerous problem Contact me at Israel120.com/contact today to learn more.

16. I want to share with you that there is hope, even in the midst of the chaos in

the world. If I, a Beninese-born American life scientist, who graduated as the doctoral marshal of my university in the US, found myself decoding complex things that physicists and chemists should have discovered centuries ago, and I published 9 popular books that are shockingly changing science forever and flipping the debate between science and faith on its head, there is nothing we cannot do if God is with us as we diligently work on our mission on this Earth in spite of obstacles we may meet. But if we turn our face against Him or start chasing things we were not meant to do, then we are in a big problem! Interview me or invite me to your organization to hear how I shockingly managed the 12 critical years that led to the publication of the 9 groundbreaking books I released in 2025!

17. More and more people are believing that it is impossible for science and faith to meet; therefore, most people are denying God at the profit of secular theories or wrong religious doctrines, which I just proved contain raw data that proves the existence of God. I want to tell the whole world why this trend of using secular science to deny God is tremendously dangerous, and how, scientifically using the Biblical account of creation, we can fix it instead of following the dangerous trend of embracing secularism while denying God because you think it is impossible for science and faith to ever meet. We cannot afford to do nothing any longer. Likewise, although there are many churches across the globe, many Christians have been unfortunately disappointed in and victimized by how their Christian belief suddenly collapsed or failed them when confronted with some secular theories. Therefore, many Christians are abandoning their faith at the profit of secular doctrines denying God's existence. For instance, more and more people are embracing evolutionism and trashing the Biblical account of creation as if the Bible lied about the 6 days of creation. Likewise, increased numbers of college students are questioning the Bible and abandoning their Christian faith while kissing academic biology and physics books that program them to believe in evolutionism, the Big Bang theory, and other theories that reject God. In contrast, many Christians think that secular science is dangerous and that science can never prove creation. Therefore, most believers are not even trying to explore how the scientific data can fit the Biblical account of creation. I am here to scientifically tell the whole world that these trends are extremely dangerous to humankind and what needs to be done to fix them. I am here to warn everybody why these tragic trends are serious and what needs to be done to quickly fix them scientifically. I am here to warn the public about these dangerous trends and prove how the historic scientific explanation of the Genesis story proves what needs to be done about this development in our modern societies. If we ignore all these warnings, anti-

creationism will prevail when Christians do nearly nothing to scientifically demonstrate creation and stand for God. Don't blame God when you see this world upside down very soon and wars blazing in nations that we once thought were unattackable and superpowers falling to the ground! But if we take these warnings and the God of Israel seriously, we will win even when our economy may seem weak, even when many problems will still exist, and even when we will be labeled by many wrong names. We cannot afford to do nothing any longer! That is why I am teaching the whole world how science accurately supports the Biblical account of creation. If we don't fix those trends, we can be found liable for the fall of our civilizations and the rise of other civilizations, which already seems to be the incoming trend, if not here yet. In other words, if some nations keep denying God, they should not be crying for the fall of their "power" while others are rising. Like it or not, the blessing of some countries and their advantage over others is because of the grace of God. And we cannot deny that God forever and continuously expects His blessing that we ignore. And someone needs to sound the alarm so people can wake up or refuse to wake up so their retribution will be inexcusable. But if we fix these dangerous trends, you can:

- avoid regrets in the end
- find new opportunities focusing on real research questions,
- expand your understanding of God's existence and market
- cut useless God's existence cost and research
- stop wasting time on useless God's existence questions that will yield nothing
- start focusing on the genuine problems of God's existence.
- spend your resources on profitable God's existence projects
- take yourself and your organization to a higher ground
- open new groundbreaking doors related to a proper testing of God's existence
- boost your profits today and forever
- properly handle God's existence debates and make eternal profits
- conserve time wasted on useless God's existence inquiries and spend it on more valuable problems
- understand your past and present, and prepare for your future with a sound knowledge of God's existence and what He wants from your own existence

18. Before I close this book, if you are a leader of an organization that can benefit from correctly understanding the demonstration of God's existence or the origin of the universe, of life, and of chemicals, I want to leave you with 24 questions to ask before hiring any expert, speaker, or consultant related to those topics. Whether you are looking at different speakers right now or in

the future, let me share with you the following questionnaire that will help you make a good decision, whether it is with me or not:

1. Does his/her organization specialize in working with experts on God's existence, the universe's origin, life's origin, and chemicals' origin?
2. Does the expert customize real content so it can be specific to our industry?
3. Do his/her topics directly match my main problems and my customers' needs concerning God's existence, the universe's origin, life's origin, and chemicals origin?
4. Is the expert enlightening, and will he/she help fix things up rather than just lecture us and go?
5. Does the expert involve and engage the audience properly during his presentation?
6. Is the program formatted to fit diverse learning styles?
7. Does the program have well-defined steps to help solve my problems?
8. Is there any online version of the program available 24/7 so I and my team can continue learning at our convenience?
9. Will I (or my team) learn new strategies to improve my (our) situation and skills quickly?
10. Will the program intellectually stimulate and challenge me (or my team) very well?
11. Is the content of the program up-to-date, specific, current, and filled with innovative information, ideas, and insights that cannot be easily found on Google or Yahoo?
12. Will I (or my team) be learning from a proven expert who is easy to work with?
13. Does the expert have verifiable client success stories that prove his/her experience and accomplishments in the real world?
14. Will there be any hidden fees or negative surprises after my team and/or I sign up for this program?
15. Has the expert been doing this job for a long time?
16. Will the expert relate to and connect with me (or my audience) very well?
17. Is he/she authentic enough to share with me (or my team) both his/her mistakes and success?
18. Does his/her company qualify clients, or does it work with anyone, meaning that his/her company is a generalist that is not specialized in a specific niche?
19. Can the author rationally demonstrate the Biblical creation without

falling into the trap of taking sides between science and faith or between reason and religion … so that I (and my organization) can be equipped to win over people from both sides of the aisle separating science and the Bible?

20. Can this expert scientifically demonstrate that the days of creation were literally 24 hours each?
21. Can the expert scientifically prove that the formation of the Earth was completed on the 3rd day of creation?
22. Can the expert scientifically demonstrate that the Moon and the Sun were formed on the 4th day of creation?
23. Can the expert scientifically prove that during the formation of the universe, layers of water existed and were separated?
24. Is the person a true scientific go-to-expert and authority on issues pertaining to the formation of the universe, of life, and of chemicals with an emphasis on turbulence and the reconciliation of the disagreements between science and creationism?

Feel free to use these questions to vet people, like me, who are seeking or may be interested in a similar speaking, mentoring, coaching, or consulting job from you or your organization concerning God's existence or the origin of the universe, of life, and of chemicals. I want to help you not just to hire me or Science180, but to make the safest hiring decision. You can learn more at www.Israel120.com/speaking

9.3. An unbelievable, historic crime story you will thank me for sharing with you

As I close this chapter, let me tell you a very touching story. Indeed, once upon a time, there was an innocent being who was accused of something He said He did not do. Unfortunately, except for that innocent being, nobody else was there when the events related to the accusation happened. Being the only witness, that innocent being told a story about what took place. He even gave a timeline and extensive explanation of the processes involved in the events that occurred. Despite the proofs that He gave, His accusers refused to believe in His story. In fact, His accusers were unable to decrypt or properly explain or understand the language that the innocent being used in His narrative. Some of those who believed in the narrative could not scientifically demonstrate it to the accusers.

That innocent being has been accused by some people who said He was lying, while He has been believed by others who could not demonstrate the things related to the history the innocent being said. Meanwhile, the accusers have managed to use the properties and resources of the innocent being to live on, to design methods to collect data to prove that innocent being wrong. They succeeded in collecting tons of data according to their mindsets, but because they failed to demonstrate that the story of the innocent being is correct, they believed in their lies, and continued to accuse the innocent being until one day, someone who believes in the innocent being but who has been tortured by some of the aforementioned accusers decided to review the data collected by the accusers to see if he could prove them wrong and prove the innocent being right.

After suffering for 12 years, that believer in the innocent being ended up discovering that the scientific data collected by the accusers demonstrated that the story of the innocent being is 100% correct, whereas the story of the accusers is 100% false. Then, the case was brought to a judge to review and pronounce a judicial verdict.

After reviewing and cross-checking all of the proofs and data collected by the accusers but analyzed by the believer of the innocent being, the judge found that the evidence confirmed that the story told by the innocent being for the past 3500 years is true indeed.

- What do you think will be the verdict of that judge? Do you think he will let the accusers go unpunished?
- Do you think the innocent being will let the accusers and all their front runners go free without paying a price for all the losses, including lives lost, because of the lies and accusations those accusers perpetuated for thousands of years?

What you just read (that came into my mind in October 2021) is a metaphor painting how unbelievers have rejected the Biblical story of creation for thousands of years and accused God and His followers of being liars, while the true liars have been the accusers. In this story, the innocent being is the God of Israel, whom people accused of claiming to be the Creator that they think He is not. The accusers are the unbelievers and even the believers who refuse to believe in the six 24-hour -day creation story of the Bible. The believer in the innocent being who demonstrated that the 3500 -year-old story of the innocent being is 100% correct and true is me, Nathanael-Israel Israel. The judge in this story is God the Creator.

Therefore, considering everything, you need to be careful about what you say and argue about creation from this point forward. Don't say I never warned you! For we are not just talking about a mere scientific thing here, but about crucial facts that have eternal implications: eternal life or death and much more, as I better detail in my book titled "*Reconciling Science and Creation Accurately.*" That book is what I called the "Biblical or prophetic version" of my book on the origin of the universe, and it targets Christians and anyone interested in knowing the Biblical perspective of the creation of the universe. That important book accurately demonstrates the marvelous creation and formation of the universe by God in six consecutive 24-hour days and answers many questions about the universe 's creation so that after acknowledging Him, you can choose and enjoy life today and forever. Get this thoughtful book now to figure out what happened at the beginning, what is coming up, and why it is time to urgently rethink everything you have been told about the universe's origin so you don't eventually regret it! Don't say I did not tell you! Learn more at

Science180.com/biblical.

While they refuse to believe in what God said, many people have died ignoring the proofs I presented in this book and the 8 others I published in 2025, which were not proven in their days but which were supposed to be only believed just as the Bible says. Instead of taking God's word seriously and simply believing He is telling us the truth, some people said there is no God. You don't want to make the same mistake as some of them and end up regretting it forever in the afterlife, when all proofs will be shown, and it will be too late to believe. For the God who created everything also talked about other things, including how people's choices in their life on Earth will affect their destination, either in heaven or in hell. Just as Noah warned the people of his generation to repent and get ready for the Great Flood, which ended up surprising people in those days (Bible's Book of Genesis 6-11), so also God has called me to build a scientific ark that He will use to redeem some stubborn scientists who (despite the proofs) refused to acknowledge God, therefore setting themselves up for hell if they do not repent. If you still do not believe in God yet, I am asking you to review the evidence I presented here and reconsider your choice, for you will be held accountable after you die one day, and you will not be able to say that nobody warned you.

Taking seriously my calling, my cross, which I initially struggled to recognize, I have finally accepted to sacrifice my plant scientist, entomologist, and microbiologist career, for which I worked very hard to earn in the USA, to help save people of all ages like you, my friend, and your dear ones. Hence, I did my best to break my findings down into books targeting different people (see www.Science180.Com/books):

1. *"Turbulent Origin of the Universe"*[19] (Tailored to scientists or people who can understand science)

2. *"From Science to Bible's Conclusions"*[20] (Targeting the general public who wants to learn the science that led to cracking the code of the universe's formation)

3. *"Reconciling Science and Creation Accurately"*[21] (Written to answer all biblical creation questions)

4. *"Turbulent Origin of Chemical Particles"*[22] (Tailored not just to chemists, but to anyone who can understand chemistry and the natural formation of chemical particles)

5. *"Turbulent Origin of Life"*[23] (Tailored toward life scientists (biologists, ecologists, etc.) and everybody else, even laypeople, who are interested in the origin of life, or those who like to study nature)

6. *"Origin of the Spiritual World"*[24] (Targeting anyone who wants to learn about ancient, hidden, and rejected secrets about the origin of everything in the universe)

7. *"How Baby Universe Was Born"*[25] (Written for secular children ages 7-12 who or whose parents don't believe in God)

8. *"How God Created Baby Universe"*[26] (Written for children ages 7-12 who believe in God)

CHAPTER 9. WHY AM I TEACHING THE WHOLE WORLD HOW SCIENCE ACCURATELY SUPPORTS FAITH AND A BIZARRE CREATION STORY … AND WHAT CAN YOU DO TO AVOID THE DANGEROUS THING THAT WILL HAPPEN NEXT?

9. *"Science180 Accurate Scientific Proof of God"*[27] (this is the book you are reading now)

10. **"Mathematical Proof of God's Existence at the Intersection of Science and Faith"**[28] (tailored to scientists and mathematicians)

If this book does not fit you, please visit www.Israel120.com/books to check out others I wrote, for some of them may satisfy you.

Why is the decoding of the discovery of the scientific demonstration of the creation story in the Bible happening now? Because the end is near! The God of Israel who gave me the grace to decrypt how He created the universe and also gave me the grace to tap into supernatural facts pointing at the end of this age very soon. In my forthcoming book on the age of the universe, I explained why I know that the end is very near and how I was able to calculate the approximate date of the end, which is within just a few years from 2025. I have a book on that coming up very soon. To be among the first people to know when that book is released, go to Israel120.com/newsletter and sign up today.

At this point, it is up to you what you want to continue to believe, but in the end, I know I accomplished my mission! In other words, to those who will continue to deny or reject God despite the proofs that I scientifically conveyed, I want to say that I at least did my part, and it is up to you what you choose to believe in, but you no longer have the excuse to say that the scientific proofs of God's existence were never presented to you. For after all, I also understand that some people will continue to argue and refuse to embrace the truth regardless of the scientific evidence before them. But I pray and wish that you will not be among those left out because of their unbelief. All I would like for you to do is to sit down, review all of the evidence I presented, and properly reconsider your decisions concerning the origin of the universe if you still deny God. To you who believe in God, I would like for you to encourage yourselves and take this message to others. For the end is very near!

With everything I've said or without accepting to risk my life so I better live later? Or were you thinking that I can demonstrate the existence of God without irritating both creationists and anti-creationists? If not, how can I differently use your doubt or belief statement as a starting point to scientifically test the existence of God without making you or those who believe in something else stop reading this book before discovering my findings and discussions? That's one of the challenges of writing a book like this. But I am glad that we all can agree on some facts nobody can change. I am also glad that evolutionists support that it is not impossible to demonstrate that God created the universe because my discoveries properly tested and proved that God did it directly

I am also glad that evolutionists support that it is not impossible to demonstrate that God created the universe because my discoveries properly tested and proved that God did it directly indeed

indeed… I am glad that, like I previously said, Positive Atheism said that, *"If anyone thinks he has a truly original argument to present to you, you will do your best to give it a fair look.* Don't you think I should be glad because I know I have a case for God's existence and I can serve as a witness, despite not being there at the beginning? But if I tell you I was there in God as a spirit before He laid the foundations of the universe, will you believe? Well, regardless of how you may rank my God's existence proofs, which questions you think I completely addressed for both creationists and anti-creationists, and which ones you think I did not address, I know that what I handled in this book is sufficient for you to accept God's existence.

As I am trying to help unbelievers to know and fall in love with God, their Creator, I cannot close this chapter without drawing your attention to some believers' blunders that—(unbelievers should not focus on?)—can roadblock you and others. For instance, can Christians succeed in massively converting freethinkers to Christianity, as they wish, if the Christians and their leaders keep shifting the message of the Cross to any of these 50 questions:

1. Who is the richest pastor or prophet?
2. Who has the biggest megachurch?
3. Who has the largest church auditorium?
4. Who has the best sound system in their church?
5. Who is the greatest minister of our time?
6. Who is the best pastor or the best preacher?
7. Who is the preacher that traveled to the most countries in the world?
8. Who is the bestseller of the most famous Christian book of the year?
9. Who wrote more Christian books this year or every year?
10. Who is the preacher or prophet that travels only in first class?
11. Who is the preacher that even has the most expensive luxury jet?
12. Who is the preacher that wears the most expensive Gucci or the most expensive brand?
13. Who is the minister that drives the most expensive car?
14. Who is the preacher that wears the most expensive watch or suits?
15. Who is the minister that owns many mansions even when his most faithful pastors are seriously struggling to put food on the table?
16. Who is the preacher who builds the largest, biggest, and most expensive city in the world while his pastors are struggling to pay rent?
17. Who is the prophet who never fails prophecy, as he knows or can know everything?
18. Who is the prophet who sees in the spirit all the time, who knows everything, and who can prophesy about everything anytime, whether his eyes are closed or open?
19. Who is the prophet who speaks to God face-to-face all the time and to whom everybody must submit?

20. Who is the preacher who does not need his church's members to pray for him?

21. Who is the preacher that can survive 21 days of dry fast, drinking no water and eating no food?

22. Who is the preacher that can take it to another level by fasting (food and drink) for 90 days?

23. Who is the preacher who can spend more time on his prayer mountain or in the forest praying, even under the rain?

24. Who is the prophet that has spiritual sons and daughters across all continents, yet himself has no spiritual father he properly bows down to as he wants others to do for him?

25. Who is the prophet or the preacher that has more international visitors than all other preachers in the world?

26. Who is the minister who make more money from this kind of tourism and who does not travel much to any country himself?

27. Who is the preacher that preaches more about sinners going to hell than God's grace capable of saving some repented sinners before they die?

28. Who is the man who is running after economic progress more than after his personal relationship with God?

29. Who is the preacher who wants to economically prosper people while disregarding his and/or his followers' spiritual state or fate?

30. Who is the person who is easily preaching Jesus to others, while he himself refuses to properly obey and live for that Jesus?

31. Who are these Christians who are fighting to be allowed to pray in public, while they refuse to pray in the secret place of their houses?

32. Who is that Pharisee that is putting heavier loads on people's heads and necks than they can carry, while he himself is following neither the law of the land nor the law of God, let's say the 10 commandments, that they want "sinners" to memorize and live by?

33. Who is that person who is bragging about God in public but who is living a personal, filthy, dirty life that people ignore but that God knows so well that he will account for one day?

34. Who is the most powerful preacher today?

35. Who is the preacher whose spiritual sons and daughters are presidents of nations?

36. Who is the preacher who has laid hands on more presidents than all other men of God?

37. Who is the preacher who has held the biggest crusade ever?

38. Who is the best preacher, pastor, or prophet whom all other preachers must follow?

39. Who is that Christian who complains more about strangers in his country than he realizes that he himself is an immigrant on this Earth that we all will leave very soon?

40. Who is the true prophet, and are all other prophets that you don't like the fake prophets?

41. Who is the preacher that is God's biological microphone that all other preachers and unbelievers must listen to?

42. Who is the preacher that has the most followers on YouTube?

43. Who is the preacher who is liked the most on social media?

44. Who is the preacher who has more followers on Instagram and TikTok?

45. Who is the preacher that has the most beautiful ties and shoes?

46. Who is the best preacher who owns a TV channel or who regularly appears on the most famous TV stations?

47. Who is the preacher who is the best top ambassador of his country—not ambassador of God—but ambassador at large of his poor yet very rich country?

48. Who is the preacher who speaks and expects everybody else to pay something, including kneeling down, to receive impartation, while he does the same to no one?

49. Who is the preacher who is waiting to raise enough money before he can go win souls?

50. Who is, who is, who is … that person … who refuses to carry his own cross but wants others to carry it … or else he will do ABC to them?

Why don't Christians actually understand the urgency of the moment and change the tone of their message and life so people who really need God can actually believe in the God of grace? For as long as Christians and their leaders keep focusing on the aforementioned nonsense, it will be very difficult for any scientific demonstration of creation to massively convert some freethinkers who are watching and grading how Christians are portraying the God they want unbelievers to accept by faith. But is it what they were supposed to do if they really know that their own life is in danger? Now that we know that they don't know, what can we do so they know that they don't know they need to know? With all these facts, do you think I can prove God's existence to humankind without angering both the believers and unbelievers?

Like I said in the beginning of this book, there are questions we will never answer. Therefore, it will benefit you to not let them force you to continue denying God if you have not started believing in Him yet. But if you want to continue to deny Him, I rest my case and let Him judge the rest! I am done. Now, let's conclude this book!

10. CAN ANYONE DEMYSTIFY THE INTERSECTION OF SCIENCE AND RELIGION TO GET SMART FREETHINKERS TO HONESTLY CELEBRATE GOD'S EXISTENCE WITH RATIONAL CREATIONISTS?

Can science and faith meet on a hot topic such as the origin of the universe? Yes. Can any creation story be scientifically demystified in the 21st century? Yes. Can freethinkers celebrate anything rational with creationists, whom they think are irrational? Yes. Does God exist? Yes. Can science prove it in a simple language that all can understand plainly? Yes. How can this rational demonstration be done quickly without fail? By intercepting science and religion impartially. Is there a simple scientific formula accurate enough to clearly point us toward the God who created the universe? Yes, and that formula is what I termed the "Turbulent Universe-Origin Formula," or the formula of the birthdate of the celestial bodies:

$$T = \frac{D}{Ve} + \frac{2\,R\,\pi}{Vo}$$

Before I go over that formula with you again, I will first remind you of some key things that led to it. Indeed, during my test of the existence of God, I discovered that both the pains and expectations of the creationists and the anti-creationists have a common denominator, yet, these people fight as if they don't have any problem or as if their opponents are the problems. I realized that countless people who have trained themselves in secular schools of thought are being left out partially because the conventional ways of sharing the creationist message does not fit their intellectual curiosity and methodology of thinking, seeking, finding, and receiving. Several scientists are unable to discover the mistakes in their scientific interpretations of the world and how it was formed, and their scientific training is unfortunately acting like a yoke that prevents them from believing in God, whom they defend cannot free them. Besides the unbelievers who reject the Bible and other religious books, many people who say they believe in God are misled, betrayed, and misguided to believe in scientific and religious lies, which apparently sound good and true to them, but which are really against the truth that even science proved. Numerous believers embrace wrong secular scientific theories and wrong creationist theories, while necking secular theories (e.g. the Big Bang theory and evolutionism) that they say oppose their religious book (e.g. Bible), yet they accuse the nonbelievers of denying God. In the end, these believers or so-called believers are like blind trying to lead other blind into a pit they ignore.

The problem is that human beings have been trying to acquire knowledge or interpret the invisible or the unknown using wrong means and then applying what they think they know to try to achieve their goals, which don't always support reality that can be hard to distinguish from fiction in our modern digital world about to be dominated by artificial intelligence.

Although some scientists didn't initially choose to deny God, the way data around them have been scientifically interpreted over time has caused them to rationally embrace theories opposing God, which even some believers are struggling to scientifically prove. Some people think that faith in God or in any deity or thing does not require physical proofs, but some people's mind is so stuck into looking for proofs for everything that it is impossible for them to accept even things that are self-proven or self-explanatory or that can never be proven scientifically. Therefore, if something is not done to help most people to know that their scientific data or spiritual dogma can correlate with a certain reality (that they rejected or embraced blindly in the name of science and/or faith), which they said they are not seeking in their ideology or philosophy, it is near impossible for most people to ever come to know the real scientific test that can prove God's existence plainly in a way that even children ages 7-12 can rationally understand like I did in some of my books (see www.Israel120.com/books). Therefore, even the most sophisticated scientific proofs about the existence of God may never satisfy everybody, no matter the number of details presented. For there are some deep expectations and pains connected to these problems that go beyond what science and even religion can ever solve or agree on.

To impartially investigate God's existence, pains, and expectations in the world to properly handle faith and doubt scientifically, I did not leave anyone out. I showed that everybody experiences certain God's existence problems. Even if they don't know, everybody takes some risks while trying to solve God's existence problems. Everybody experiences some emotion or encounters annoyance while trying to solve problems related to God. Finally, everybody is hoping to gain something from most efforts related to testing or proving God's existence or inexistence.

I showed that, although they may not admit it, people of all backgrounds and belief systems struggle with the scientific demonstration in favor of or against God's existence, at least in the eyes of their opponents. After reviewing these pains and expectations, I explained my objective for this book: scientifically scrutinize the existence of God using all existing data without falling into the trap of taking sides between science and faith. Toward that end, I proved that creationists need to take seriously the criticism of freethinkers and rationalists, and vice versa. I also proved that Christians are not as irrational as most secular rationalists and freethinkers think they are.

Then, I explained how I discovered the "Universe Turbulent Origin Formula" that you are not using yet but that is the accurate key to scientifically proving God's existence. And how do I know this formula is correct?

Telling you my story, which started in a small country called the Republic of Benin in Africa, I explained how during and after my education that culminated with the obtention of my PhD in plant, insect, and microbial sciences at one of the best universities in the USA, I did not know I was going to write this book. But a concourse of circumstances caused me to write and publish 9 books on the origin of the universe, of life, and of chemicals in 2025.

I did not use the conventional scientific method, which usually consists of first reviewing the literature and then laying hypotheses, which are then tested after designing a method of research to collect data, which are analyzed so that rational conclusions could be deduced. But in the case of this study, everything started with ideas coming to my mind after I asked certain key questions about life and how to make it following certain experiences in Africa and in the USA. These ideas landed on the formation of the universe.

As I decided to check the literature to understand what science said concerning the ideas I was getting about the celestial bodies, I landed on a gold mine of scientific data collected and trashed by top scientists, including those at NASA, but they, and other scientists across the globe and throughout the ages, did not value those data, else, they would have focused their attention on analyzing them differently rather than investing billions of dollars into asking and researching useless questions. concerning the origin of the universe, of life, of chemicals, and the existence of God. Why am I saying this?

As I started viewing these data, trends were appearing to me at a very fast speed. But when I searched the literature, how these data were analyzed and the conclusions drawn from them were different from the interpretation I was getting for them. As I continued reviewing these data, I was getting so many ideas that I could not ignore. I then realized that I had a story to tell. Over 12 years, the study and analysis of these data led me to write 9 amazing books targeting different aspects of the formation of the universe, of life, and of chemicals. I tailored these books not only to scientists of all kinds, but also to laypeople and even children (see www.Science180.com/books).

When I was analyzing the data, I came to a point where I had a system of equations I needed to solve. I knew there must be a certain phenomenon behind the trends of these equations, but I did not initially know what that phenomenon was. After working on the data and researching the root cause of the trends I was getting, I discovered that the trends in the data were caused by a phenomenon called turbulence, an unsolved mystery in science. I had to design a methodology of research to analyze the data on the celestial bodies in the perspective of turbulence.

During that process, the scientific data suggested that, at the beginning of the universe, a certain initial matter, which I called the "turbulent prima materia" mysteriously appeared over a large portion of space. As this material was destabilized, it was split into the precursors of daughter bodies which, at their turn, were also destabilized and reorganized into the precursors of various celestial bodies following a cascade of split and gathering of what I termed precursors of bodies. I extensively explained that process in my book called *"Turbulent Origin of the Universe."*

The scientific analysis of the data showed that, even the Earth, the Moon, the Sun, and every celestial body in the Solar System had a precursor. The science indicated that, during the formation of the universe, the precursor of the Solar System birthed 2 precursors:

- The precursor of the Sun (which went through a process to become the Sun), and
- The precursor of the bodies orbiting the Sun, which birthed all of the celestial bodies orbiting the Sun

A difficult task of this study was the scientific explanation of the process that reorganized the initial matter in the universe into the celestial bodies we have today. To address this challenge, I had to study many variables collected on various celestial bodies in the universe. Of the hundreds of variables that I analyzed, 4 stood out the most:

- Escape velocity (which is the minimum speed needed for a celestial body to escape the gravity of another one)
- The semi-major axis (which is the average distance separating a celestial body from its primary body, meaning the body it orbits)
- The orbital speed (which is the speed with which a celestial body orbits another one)
- The radius (which is the average distance from the center of a body to its periphery)

While working on this book, I realized that the aforementioned 3 celestial bodies (the Earth, the Moon, and the Sun) and these 4 simple variables were sufficient to demonstrate how the universe (including the Solar System) was   formed and to quickly, easily, and scientifically test God's existence.

As I was reviewing the literature, some key questions I asked, which now serve as the cornerstones of my scientific proofs of God, are

1. Why is Mercury (the closest planet to the Sun) located at 57,910,000 km from the Sun?
2. Why is the distance separating the Earth and the Sun 149,600,000 km?
3. Why is the distance separating the Moon and the Earth 384,400 km?
4. Why is the orbital speed of the Sun 19.4 km/s?

5. Why is the Earth orbiting the Sun at 29.78 km/s?
6. Why is the Moon orbiting the Earth at 1.02 km/s?
7. Why is the radius of the Sun 696,000 km?
8. Why is the radius of the Earth 6,378.14 km?
9. Why is the radius of Moon 1,738.10 km?
10. Why is the escape velocity of the Sun 617.6 km/s?
11. What is the escape velocity of the Earth 11.186 km/s?
12. Which religion has a creation story matching these scientific evidences?

Formulated by me, Nathanael-Israel Israel, these questions are known as "The 12 Universal Equations of God's Existence." If you can correctly answer these questions, you can properly explain the origin of the universe and scientifically prove the existence of God.

I discovered that the celestial bodies we have today did not come out of one another, but, as their mothers or precursors broke, they were reorganized into systems of bodies according to a rule that I summarized in the concept I termed "split-gathering," a process during which precursors of bodies were split and gathered along a cascade until all of the bodies in the system were born. I invented the term "split-gathering" to express the processes by which fluid layers in the early universe were split from one another and then gathered together into bodies that we see today. In general, this process of split-gathering can be summarized as consisting of a mother body that split into daughter bodies, which then traveled a certain distance before being wrapped around to form adult bodies. Sometimes this process can repeat itself over a cascade of split-gathering before the daughter bodies could be finally formed. I used the term "precursor" to qualify the body that was split-gathered together to birth another one. In other words, the precursor of a celestial body is the body that was split-gathered to birth that celestial body.

I showed that the precursors of bodies were made of certain particles that were not defined yet, and these particles went through changes before becoming what we may call "Adult particles" today. In my book *Turbulent Origin of Chemical Particles*," I scientifically explained how the formation of the celestial bodies and of the chemical particles they contain took place at the same time. In other words, chemical particles were not formed before or after the formation of the celestial bodies that contain them or between which they are found. But as astronomical fluid layers were being split-gathered into celestial bodies, the matter in them, which at one point was fluid or fluid-like, was organized into simple or complex particles according to their size, position, and movement. I showed that the way the fluid layers were split-gathered is like a gigantic seed that germinates and grows to a point before birthing branches bearing leaves, until fruits can appear before the growth stops at one point. In other words, I showed the fluid layers of the precursors of celestial bodies moved and split into different fluid stacks until the last fluid stack was released.

I demonstrated that the fluid layers on top separated first, while the fluid layers at the bottom were the last to split from the rest. Just as the branches of plants are not positioned at the same point, so also the daughter bodies of a mother precursor were not born at the same time but at different times according to the duration of the processes that birthed them. I observed these processes not only with celestial bodies but also with chemical particles and even with living things. Although they share some similarities, these processes are expressed with some nuances according to the precursors involved and whether it is a living or non-living thing. In my book called "*How Baby Universe Was* Born," I plainly explained these processes in a language that even children ages 7-12 could easily understand. In my book "*Turbulent Origin of the Universe*," I spent hundreds of pages detailing this process in scientific language.

I showed that, in the Solar System for instance, the precursor of the bodies orbiting the Sun escaped the precursor of the Sun at about the escape velocity of the Sun and then started moving away from the precursor of the Sun. I scientifically proved that by 26.05 hours after the beginning of the formation of the universe, (that's on the second day), the precursor of the bodies orbiting the Sun split from the precursor of the Sun, and its fluid layers started splitting from one another (see Chapter 4). This finding agrees with the Biblical account of creation, which mentioned that waters started to be divided from waters on the second day of creation.

I showed that, after the precursor of the bodies orbiting the Sun escaped the precursor of the Sun at about 617.6 km/s (i.e., the Sun's escape velocity), it was organized as a stack of fluid layers in which the precursor of the Earth-Moon system was embedded and had to "wait" until all of the fluid layers of the bodies located above it (e.g., the precursor of Mercury, of Venus, and of many asteroids located between the Sun and the Earth) split and were removed before it could (at its turn) split-gather into the precursor of the Earth and the precursor of the Moon.

I showed that the distance traveled by the precursor of the Earth-Moon system before it split from the fluid layers below it was about 149,600,000 km (i.e., the semi-major axis of the Earth). Dividing that distance by the escape velocity of the Sun (i.e., 617.6 km/s), I proved that the time it took for the precursor of the Earth-Moon system to reach the position of the Earth was about 67.29 hours (i.e., 2.804 days).

The precursor of the Earth-Moon system split quickly, yielding the precursor of the Earth, which was then wrapped around at about the orbital speed of the Earth (29.78 km/s) to form the Earth, which has an equatorial radius of 6378.137 km. I showed that in 22.42 minutes (i.e., Earth's circumference divided by Earth's orbital speed), the fluid layers of the precursor of the Earth were gathered together.

Adding together the time it took for the fluid layers of the precursor of the Earth-Moon system to move from about the precursor of the Sun to the orbit of the Earth (67.29 hours) and the time it took to be gathered together around its circumference (22.42 minutes), the duration of the formation of the Earth is:

Earth's birthdate = 67.29 hours + 22.42 minutes = 67.66 hours = 2.82 days

This agrees with the Bible's Book of Genesis, which said that the formation of the Earth was completed on the 3rd day of creation (Genesis 1:9-13).

I also proved that the precursor of the Moon split from or escaped the precursor of the Earth at about the escape velocity of the Earth (11.186 km/s) and traveled for about 384,400 km (semi -major axis of the Moon) before reaching a point where it was collected into the Moon. By dividing the semi-major axis of the Moon by the escape velocity of the Earth (384,400 km divided by 11.186 km/s), I showed that, after traveling for 9.54 hours away from the precursor of the Earth, the precursor of the Moon, was ready to collect itself into a satellite.

384,400 km divided by 11.186 km/s = 9.54 hours

I demonstrated that the time it could have taken for the fluid layers of the precursor of the Moon to gather themselves into the spherical Moon (radius = 1738.7 km) after it reached its semi major axis was about the circumference of the Moon divided by the orbital speed of the Moon:

2 * 3.14* 1738.1 km divided by 1.02316 km/s = 2.96 hours

Therefore, the time elapsed before the Moon was fully formed is the sum of:
- the time it took for the precursor of the Moon to escape the precursor of the Earth-Moon system and reach the orbit of the Moon (9.54 hours) and
- the time it took for the fluid layers of the precursor of the Moon to be wrapped around to form a spherical body having a radius like that of the Moon (2.96 hours):

Moon's birthdate with respect to the Earth = 9.54 hours + 2.96 hours = 12.5 hours

Because it took 67.286 hours for the precursor of the Earth-Moon system to move from the precursor of the Sun and reach a position in space where it split from the stack of fluids and then split into its daughter bodies (the precursor of the Earth and that of the Moon), the duration of the formation of the Moon with respect to the Sun must consider all of the time that occurred before the precursor of the Moon was even born:

12.5 hours + 67.286 hours = 79.786 hours = 3.324 days

This means that the Moon was formed on the 4th day, just like the Bible stated more than 3500 years ago (Genesis 1:14-19).

I showed that before the precursor of the Sun started gathering its fluids into a spherical body, the precursor of the bodies orbiting the Sun traveled a distance no higher than 57,910,000 km, the semi-major axis of Mercury (the innermost planet orbiting the Sun). I proved that the time it took for the fluid layers of the bodies orbiting the Sun to escape the precursor of the Sun is about:

57,910,000 km divided by 617.6 km/s = 26.046 hours

Once the precursor of the bodies orbiting the Sun escaped the precursor of the Sun, the fluids of the latter moved at about the current orbital speed of the Sun (i.e., 19.4 km/s) and were gathered together to form the Sun, which has a radius of 696,000 km. I calculated the circumference of the Sun (2 * 3.14 * 696,000 km = 4,370,880 km), which is about the distance a fluid parcel could have taken to complete one turn around the Sun.

Using lessons from my groundbreaking discovery on turbulence, I divided the circumference of the Sun by its orbital speed to get the time it could have taken for the precursor of the Sun to swirl and form the Sun:

4,370,880 km divided by 19.4 km/s = 62.58 hours

The total amount of time it took for the Sun to form is the sum of:

- the time it took for the precursor of all the bodies orbiting it to clear the way or to escape the precursor of the Sun (26.046 hours) and
- the time (62.58 hours) it took for the fluid layers of the precursor of the Sun to be wrapped around to form a spherical body whose radius is equal to that of the Sun:

Sun's birthdate = 26.046 hours + 62.58 hours = 88.63 hours = 3.693 days

In other words, for the first time in history, I (Nathanael-Israel Israel) scientifically showed that the Sun was formed on the 4th day since the beginning just as the Biblical creation narrative states (Genesis 1:14-19).

These scientific evidence confirmed that the Biblical creation days were 24 hours each, NOT millions of years as some people have alleged. To make a long story short, among all of the religious narratives of the creation of the universe including that of Animism, Buddhism, Confucianism, Evolutionism, Hinduism, and Islam, only the Judeo-Christian version recounted in the Bible is backed 100% by the scientific data. In the Biblical story of creation, Moses clearly stated that God created the universe. Because the Biblical creation story told millennia ago is scientifically found to be true today, we must also accept the fact that Moses said that God is the Creator.

Furthermore, in addition to the Book of Genesis, Moses authored 4 other books (Exodus, Leviticus, Numbers and Deuteronomy). In those books, Moses revealed on many occasions that there is only one God, the God of Israel, referred to in the Biblical creation story as Elohim, but known to the Jews by many other names: Adonai, El, Yah, Yahweh, Jehovah, etc. In other words, although people across the globe and throughout history tend to label their idols as "gods," there is only one true God, the God of Israel, who created the heavens and earth and everything within and between them. To put it another way, although there are many "gods" or "idols," there is ONLY ONE GOD, the God of Abraham, Isaac, and Jacob.

Do you have to embrace evolution or deny Biblical creation to scientifically prove that God created the universe in 6 literal days? The answer is No. I explained why the ancient Greek thinkers born about a thousand years after Moses missed the formula that scientifically leads to God's existence. I also explained how and why Moses who lived thousands of years before the scientific age accurately predicted the Biblical creation narrative. I told you that I wrote 2 other books called *"Reconciling Science and Creation Accurately"* and *"Origin of the Spiritual World"*, in which I better elaborated on many questions I cannot fully address in this book due to its objective and space constraints. You can also join Science180 Academy (www.Science180Academy.com) to learn more.

I explained why the Islamic creation narrative is not only different from the Biblical creation narrative but also plainly wrong, as it contradicts the scientific evidence that confirmed the Biblical account of creation.

I invented the term "Science180 creationism" to label my scientific demonstration of the perfect match between science and the Bible, which led to the scientific proof of God's existence. I detailed for you why Science180 creationism's argument of God's existence is unique and different from all existing God's theories, including creationist theories. For instance, I showed that it is the:

- Undeniable scientific argument that reconciled science and the Biblical account of creation
- Easy-to-understand proof of God's existence
- State-of-the-art decoding experience of God's existence that helps you fight wasteful programs and theories
- The most accurate, reliable, safest, best explanation of God's existence ever
- God's existence rule breaker you love

To illustrate, I showed that, unlike most creationists who usually start with religious texts to make their case for the existence of God or to show how science can explain their belief, and unlike the proponents of Intelligent Design who usually start with natural empirical results to draw intelligent inferences from the evidence, not to prove God, but an intelligent designer, Science180 creationism is the first and only argument of God's existence that started the demonstration of the origin of the universe with scientific data collected by renowned secular scientists and then landed on the scientific proof of the creation of the universe by God as summarized in the Biblical account of creation.

With Science180 creationism, you don't have to throw away the scientific and rational quality of your mind to receive God or to believe in the Bible. Science180 creationism proved beyond doubt that you can know God and believe in His Word by using your scientific and rational mind 100%. Science180 creationism rejects the allegorical interpretation of the Genesis account of creation and that the days of creation cannot be viewed as allegorical. Science180 creationism proved that each creation day doesn't symbolically represent 1000 years of the world's history. Unlike what Day-age creationists advocate, Science180 creationism proved that each day of creation was not longer than 24 hours. Contrasting gap creationism, Science180 proved that there is no gap of time between the days of creation or between the first verse and the second verse of the Genesis story in the Bible. Opposing Progressive creationism, Science180 creationism showed that God did not create the universe gradually over hundreds of millions of years. Unlike Old Earth creationism, Science180 creationism proved that the Biblical account of creation is a strict chronological record of how God created the universe and its content. Unlike Old Earth creationism, Science180 creationism exposed that the Genesis story is not a topical order of events. Opposing Day-age creationism and all other types of Old Earth creationism, Science180 creationism scientifically demonstrated that the days of creation were ordinary 24-hours consecutive days that do not represent *"much longer periods of millions or billions of* years" is not part of an *"infinite antiquity"* that ended when the Spirit of God moved upon the face of the waters.

Because Science180 creationism explanations can be tested, modified, or rejected by scientific means, they can be included in the processes of science. Unlike Intelligent Design, Science180 creationism does not point at the creator of the universe as a simple designer or engineer but engineer, but as the God of Israel, the God of the Bible who spoke to Moses, the major prophet who wrote the Biblical account of creation in the Bible's book of Genesis about 3500 years ago. Science180 creationism proved that, just as the first verse of Genesis says, God created the Heaven and Earth in the beginning, yet it took some time before the Earth could get its final form (which was completed on the 3rd day), so also at the end of the 6 days of creation, everything in the universe was created, but it took a certain additional time for some of them to be fully formed and get their final shape. For instance, Science180 creationism, which scientifically proved that the Earth was formed on the 3rd day, while the Moon and the Sun were formed on the 4th day just as the Bible says, also proved that all of the planets in the Solar System were not formed yet at the end of the 6th day, but their precursors had to go through some processes (which took a few more days) before they were fully formed (details for the birthdate of celestial bodies can be seen in my book titled *"Turbulent Origin of the Universe"*).

Unlike Young Earth creationism, which has a better perspective than all existing creationism theories before the days of Nathanael-Israel Israel, but never actually demonstrated the Biblical creation scientifically, Science180 creationism (spearheaded by Nathanael-Israel Israel) actually properly demonstrated the Biblical creation to a scientific standard that considered the real needs of freethinkers, therefore making countless freethinkers to fall in love with the God of Israel they have denied. In short, for the first time in history, Science180 creationism exploded the myth of those who think that the days of creation in the Bible were millions of years instead of 24-hour days each. It also scientifically challenges and refutes the belief of the Big Bang theory and all other theories that think that there is no God who created the universe. It proved that the God of Israel is the one who created the universe. It also suggests that this proof is not an endorsement of everything the nation of Israel has been doing, for, although the Messiah is from Israel, most of the Israelites (except the Messianic Jews) have rejected God, even until today. Yet, the nation of Israel is the apple of God's eye, the timetable of the world that informed people should look at to know where we are in the history of the world: the end is near. I addressed that topic in other incoming books (learn more at www.Israel120.com/books and to see where we are headed).

Science180 creationism also highlights the message of the grace of God, without which no one could be saved, not even the popular preachers or prophets who, just like some unbelievers, have made gigantic mistakes about the Biblical creation and God's existence to the point of embracing anti-creationist theories while holding the Bible in their hands, even while in the pulpit. Now, let's finish the conclusion with the answer to some questions I asked earlier.

Can anyone demystify the intersection of science and religion to get smart freethinkers to honestly celebrate God's existence with rational creationists? The answer is yes. And who did it? I, Nathanael-Israel Israel, am privileged to be the first person in history to scientifically demonstrate the intersection of science and the Bible. I am grateful to God, without whom no one could ever scientifically prove His existence.

I understood that my journey prepared me for this discovery, and I, as a dual citizen of Benin Republic (a French -speaking country in West Africa) and of the USA, understood that I may not have discovered the scientific proof of God if I did not go through certain things in life, which caused me to ask certain questions that most people don't conventionally ask.

But can you scientifically prove the existence of God to atheists and all freethinkers without making the believers unhappy? The answer is NO. Why? Because, among many things, most believers throughout history have not understood the real origin of the universe. Some people believe in the right thing, but they did not know how to demonstrate it scientifically. Some people say they believe in God, yet they don't believe in the story about the creation of the universe. Moreover, I understood that some unbelievers did not just deny God, but they didn't and don't know how to rationally demonstrate the creation story of the universe. By the time I demonstrated the creation of the universe, although I knew that many people would be happy, I also understood that countless people, including believers and even devoted creationists, would be unhappy, for my discovery shines light on their wrong theories, which they may not want to quickly abandon because of many reasons, including their ego and the so-called benefits they get from them. Nevertheless, I understood that this discovery will help countless people, including freethinkers, for it rationally demonstrates what they were looking for.

My discovery suggests that rationalists and all other freethinkers cannot address the believers as if their faith is scientifically useless, while the creationists cannot continue to judge the rationalists as God's haters and dangerous, yet expect to find the support needed to advance science for the benefit of humankind. Likewise, creationists and the anti-creationists cannot solve their God's existence problems without empathically caring for the pains of their opponents and avoiding quickly falling into the trap of unconsciously taking sides in the battle between science and faith. Similarly, we cannot holistically solve the problem of God's existence by forcing all believers to drop their faith to please and kiss secular science, while asking the rationalists and all other freethinkers to deny science and embrace faith blindly as the solution to their scientific problems or to the problems in the world, which the atheists also use to deny God. I showed that disasters, suffering, evils, and pestilence almost everywhere on Earth cannot be used as a reason to disprove the existence of the all-powerful, smartest, and mighty God. In other words, using the major problems in the world is not the greatest scientific proof against the existence of God.

Getting help from math to scientifically see if any scientific theory can accurately lead us to faith, or vice versa, I proved that the Biblical creation of the universe really links science and the Bible to God. Can science lead to any religious truth about the origin of the universe, or can any religious story of creation have a scientific power that can help us test God face-to-face? Yes, science did. Indeed, as I scientifically reviewed the religions across the world, I proved that only the Judeo-Christian religion talks about the real God, yet many Christians and Jews are denying the God they (think they) are serving. Hence, some of these people are doing evil things that could not justify the existence of God to their opponents, who are failing to discern things beyond the natural that science cannot understand if we cannot use faith.

Can we really explain the formation of the universe through natural processes without evoking evolution and the Big Bang? YES!

Is there any God that smart atheists, agnostics, rationalists, and other freethinkers can actually like using reason and scientific facts? Yes, and this discovery proved that He is the God of Israel that created the universe, but most Jews and other cultures are not really following, since they don't believe in Jesus, the Yeshua HaMashiach that the non-messianic Jews killed about 2000 years ago, ignoring that God planned all that so He could save humankind.

Do people, including most fervent creationists, believe lies about creation that need to change so atheists and all other freethinkers can enjoy the God they hate? Yes, and those lies can be summarized into all of the creationist theories that oppose the literalism or inerrancy of the Biblical account of creation. But unfortunately, until my discovery, even people like Young Earth creationists, who believe in the literalism of the Genesis story, don't know how to demonstrate it, and most of them have used methods that the freethinkers labeled as prescience. To some extent, considering the demonstrations I did, I understand that most Christians have failed to scientifically demonstrate creation, and it could have even been better that some of them just defend their faith in the scriptures without forging weak "scientific" reasoning to prove what they believe. To avoid confusing rationalists, creationists need now to correct all of the existing creationist nonsense, which abounds in the variants of Old Earth creationism and all creationist theories that deny God as the one who formed the universe literally as the Biblical book of Genesis states.

Therefore, do freethinkers really hate God? Although some do, I realized that most of them just don't know the God that they deny. And throughout history, most freethinkers were seeking the scientific demonstration of something that could not be demonstrated in those days, for, among other things, God did not primarily intend for human beings to know him by primarily using science, which is limited, but by using faith, which allows believers to access forms of knowledge that are true but could not always be scientifically demonstrated yet. We are privileged to be living in a time like this when science is helping us to rationally demonstrate what God said about Himself and how He created the universe.

In other words, many believers deny God more than they think, even when they consider unbelievers as the only God doubters. Verily, some people are God believers who deny God. The demonstration I did in this book can be a starting point for the specific thing smart people can scientifically do to better know or test the inerrancy of the creation story while keeping their reason intact and free. As we all do that, we will discover that there is a real science behind the Biblical creation narrative that can practically, scientifically benefit humankind. As they understand and embrace the truth in the Biblical account of creation, smart freethinkers can actually end up liking the God that they could not prove exists but that they now know exists. In other words, if there is any scientific story that atheists, evolutionists, agnostics, rationalists, humanists, and other freethinkers can properly scrutinize and automatically rethink their argument against the existence of the God they think doesn't exist or is dead, the story in this book is it. For science teaches crucial things about creation and the existence of God that most believers, atheists, and freethinkers ignore. But from now on, this book overturned that myth of the war between science and the Bible. The battle is finished.

Some people have asked that if God exists, why doesn't he show himself? My discovery proved that God showed himself, and we just did not know, or some people refused to believe in what He said, which they could not demonstrate scientifically. What I proved in this book is part of the amazing scientific formula or discovery that can actually help atheists to love God forever. Can there be any other proof more than this?

I know that after the publication of my 9 books, a lot of scientific discoveries will be made. Some people may not even finish reading the 9 groundbreaking books I wrote before starting to write their own, maybe even claiming credit for things others have toiled for. As the knowledge of the glory of God fills the earth, unspoken discoveries will be made and leave people astonished. Crazy scientific formulas and programs will be discovered not only about the vast universe, but also about living things and chemicals particles to a detail never thought before.

However, if you are looking for the scientific proof of the creation of the universe by God, this book and the 8 others I published in 2025 showed you the formula. If this formula does not satisfy you (I hope it does), then no other formula will probably ever satisfy you. For the scientific data in the Biblical creation cannot lead to anywhere else with this math besides the existence of the God of Israel.

No other religion has a creation story that matches the science. Buddhism did not elaborate on the formation of the universe. Hinduism does not have a unique creation message. Confucianism did not elaborate on the origin of the universe. Although Islam has a creationist story of the formation of the universe, its message did not match the math or science I showed in this book and others I wrote on the formation of the universe, and of life, and of chemicals. In other words, for the first time in history, the Biblical account of creation is scientifically proven correct.

I hope and pray that my findings can open the door to the dialog between science and religion in a way that can allow Christians and freethinkers to legally work together and quickly allow the teaching of Science180 creationism in public schools instead of evolutionism and the Big Bang theory only, which were used to take Biblical creationism to the court exactly 100 years ago. I hope that this can be done in a friendly manner that will not necessarily force Christians to take evolutionism to the US Supreme Court to ban evolutionism and the Big Bang Theory from public schools.

I also hope that some Christians will finally understand that what they were calling science is a kind of pseudoscience that they need to abandon to embrace Science180 creationism so freethinkers can better listen to them and enjoy God. As that is happening, Old Earth creationists need to abandon their evolutionary love story. As they do that, atheists need to understand that God exists and that, just as your suffering today is not sufficient to allege that your ancestors or great -grandparents never existed, so also, atheists and all freethinkers cannot continue to use the suffering of people in the world to claim that God, the creator, does not exist.

It may also be helpful if some creationists can "apologize" to some freethinkers so the latter can finally listen to the message of salvation in Jesus, at least rationally if they are still struggling with faith. Likewise, Christians leaders, including political leaders, need to avoid using the discovery of Science180 creationism to crush the unbelievers. They should not use my findings to justify any vengeance toward the secularists who have casted God, the Creator, out of the public systems for a long time. For if God is happy with my discoveries, and I know He is, He may not be completely really mad at everything in secular science, which collected part of the scientific raw data I used to prove His existence.

It is possible that some top leaders use executive orders to force the teaching of

CHAPTER 10. CAN ANYONE DEMYSTIFY THE INTERSECTION OF SCIENCE
AND RELIGION TO GET SMART FREETHINKERS TO HONESTLY CELEBRATE
GOD'S EXISTENCE WITH RATIONAL CREATIONISTS?

Science180 creationism in public schools to honor God and try to promote the Bible,
but even though that may happen, we need to show love toward the lost and the
needy rather than hating them or chasing them away from the real discussions
targeting the needs they ignore.

Similarly, anti-creationists cannot expect creationists to ignore Science180 proofs
of God's existence and just continue to let evolutionism dominate, dictate, and rule
the curricula in public schools, while God the creator is cast out. With the proofs of
God's existence I provided in this book and others I wrote, I think it will be fair to
use Science180 creationism to reopen science classrooms to discussions of God, the
creator. For humankind will benefit from politicians, the judicial systems, school
boards, policy makers, and administrators finally allowing the teaching of Science180
creationism in public schools and helping embody it properly in doctrines, regulations,
and even political laws so God can bless our land.

From the Old Earth creationists to the Young Earth creationists passing by
Intelligent Design proponents, all believers must unite behind this unique effort to
honor God the creator so unbelievers can fall in love with Him in the perilous times
we are living in. Else, the future looks bad for those who will continue to oppose God
and his agenda for humankind in our time.

At the same time, the rationalists need to understand that rationalism is limited
and that there are other types of knowledge that go beyond freethinking. Your
perspective will always affect your methods of analyzing things, and if you use your
perspective to reject a fact before you start trying to demonstrate it, you will probably
not get to the truth but go where you want to go, which can be everything else except
the real destination you already excluded before the start. For there is a certain world
beyond your knowledge and even far beyond what your brain could ever understand
or imagine. And if you are not careful about how you absorb knowledge, you may
end up just swallowing distorted facts. For many scientists got a lot of data on the
universe right, but they then distorted the story of the formation of the universe as
they pleased, because their thinking misled them! You better be careful of what you
are doing with the facts you are learning or getting from nature.

Considering the evidence, I would like for you to seriously take what I said in this
book about creation and reconsider your ways concerning who you believe is the
creator of the world and how to obey and serve Him. For a day is coming when
everybody will give an account of their own life and belief. I hope you will be among
those who embrace the truth until the end of this world, which some people deny.
Just as some people have rejected the 6-day of creation story for thousands of years
before I was able to recently demonstrate it, I hope that you may come to know the
truth and believe in it now, rather than rejecting and being shocked in the life after
this one. Belief in the true knowledge is better than believing in scientific theories
based on shaky and faulty foundations.

As you embrace the truth at the intersection of science and the Bible, you will have
peace of mind, even if all your God's existence questions are not precisely answered,

for you now know there is God. By doing so, you will protect yourself and dear ones against biases, crashes, and harms due to errors and misinterpretation, therefore significantly making your efforts to scientifically understand God's existence safer and long-lasting using scientific facts and tested proofs. I encourage you to hang on to the unique perspective of how to probe God's existence, with capabilities exceeding that of any other God's existence theory, that this book gave you. This will provide you with advanced tools to explain complicated tasks as you think about or work on the cosmos and its content or as you scientifically investigate the accurate and effective test of God's existence and continue your journey on this earth, which is just a preparation for what is to come after death.

In the remaining part of this book, I have some additional resources I want to share with you to help you in our journey together and show you how you can make the most out of my discovery and even how you can partner with me. I long to hear back from you, and you can contact me at www.Israel120.com, where you can also discover other products and services I offer that can help you. If you also want to donate to help me to take this message to the whole world, you can also do that on Israel120.com/donate. I thank you for taking your precious time to read this book. Until I hear back from you very soon, be safe and careful with what I shared with you in this book. Does God exist? Yes. Can we demonstrate it using pure science and the Bible? YES, and I just did it. Thank you.

NEXT STEPS OF THE JOURNEY

Get free resources on Science180.com

If you have finished reading this book and would like to learn more about my discoveries and how they can help you, you are at the right place. Indeed, I am really committed to helping you address any questions that you may still have concerning the origin, functioning, and fate of the universe, and how you can partner or collaborate with me for greater results.

To get free resources that will help you understand other aspects of the universe's formation not covered in this book, visit Science180.com and my personal website, Israel120.com. On those sites, I will be sharing guides and strategies to get the most out of my initiatives. I will also be sharing my favorite references, tips, next-step readings, and other important things in the pipeline that will help you regardless of your field of expertise, interest, and needs.

Subscribe to "Science180 Newsletter": The only accurate newsletter on the universe's origin, life's origin, and chemicals-origin newsletter in the whole world!

Be a part of decoding the universe's origin, life's origin, and chemicals' origin! Get origin-related news, information, discoveries, updates, announcements, reviews, articles, educational materials, and opportunities from a holistic perspective not available anywhere else so you can participate in and enjoy decoding the origin, current state, and fate of the universe and its content. You will also receive priceless tips about how Nathanael-Israel thinks, what his secrets and initiatives are, what he has accomplished, and what he recommends. Without any delay, sign up for the Science180 Newsletter today at Science180.com/newsletter. It is free!

Speaking engagement

In addition to writing groundbreaking books and engaging in other business endeavors, Nathanael-Israel Israel is a renowned speaker, who you can invite to speak at your organization.

Values that Dr. Nathanael-Israel Israel can add to your life include:

- Rare expertise and tips that will increase your abilities
- Usefulness that will advance your impact regardless of your field of expertise
- Understanding of the world that will sharpen your perspective
- Critical information that will positively change your life
- Experiences turned into insight that will motivate and guide you
- Irrefutable scientific proofs of the existence of God that will save you time and launch you into a zone of unlimited opportunities
- Unquestionable scientific proofs of how God created the universe
- Accurate demonstration of the historic formula that reconciled science and the Bible
- Enlightenment that will help people including Christians to start using their brain instead of just praying and expecting God to do everything for them

For speaking inquiries, including how you can get Dr. Nathanael-Israel Israel to speak at your organization or at an event, visit Science180.com/speaking for more details.

As the standout scientific authority who accurately decoded the universe, Nathanael-Israel Israel has been helping countless people across the globe to discover and understand the complex origin of the universe without leaving out the challenging questions that people of all ages have been struggling to answer for thousands of years! As the true go-to expert when it comes to the formation of the universe and of life, Nathanael-Israel believes that, regardless of age, background, culture, religion, or profession, everyone deserves to understand how the universe and life were formed and how they can leverage that knowledge to improve lives nonstop. Therefore, his groundbreaking discoveries of the formation of the universe, life, and chemicals have been broken down into books tailored to scientists (including physicists, chemists, biologists, and mathematicians), laypeople or the general public, believers and freethinkers, philosophers, children, etc., therefore maximizing the benefits to humanity. These historic, internationally acclaimed origin books can be found at www.Science180.com/books:

When you hire Nathanael-Israel Israel to speak at your organization, you will:

- Get specific, in-depth knowledge, up-to-the-minute information, ideas, and insights about the universe's origin, life's origin, and chemicals' origin so that you expand your market, cut useless costs, stop wasting time on inadequate projects, and start focusing on the profitable solutions
- get relevant universe-origin stories that are specific to your field of expertise
- learn from a cooperative, flexible, and easy-to-work-with expert who will respond to your universe formation needs and position you to stay on top of your competitors

- interact with a renowned expert that will not just lecture you, but that will help you sort out your origin-related questions using strategies to tap into deep secrets you ignore
- listen to an experienced expert who discovered outstanding secrets about the origin of all there is
- learn authentic information not from someone who just reads you a PowerPoint, but from the true go-to expert (when it comes to critical cosmological problems) who will share with you both his mistakes and successes that will help you get much closer to the better life you want to live
- revolutionize every origin-related domain with your accurate understanding of the universe's origin.
- scientifically learn how the Earth was formed on the 3rd day of creation
- logically learn how the Sun and the Moon were formed on the 4th day of creation
- hear Dr. Nathanael-Israel Israel's personal selection and teaching of key topics that will help you break the code of the universe's formation and functioning, and strategically enlighten you, guide you to navigate and filter the massive data collected on the universe and its content so you know how to answer the world's most challenging origin questions, remove any scientific and philosophical cataracts that may be blocking you, and help bring you many steps closer to your best life today and forever
- hear the greatest scientific and philosophic lessons of some top scientists, philosophers, thinkers, and public figures who have realized historic mistakes they made in life (concerning the origin of the universe, life, and chemicals), and that they corrected thanks to the discoveries of Nathanael-Israel Israel, who founded Science180, and who is acknowledged as the scientist that truly decrypted the universe's origin for the first time
- Get world key lessons successful people have learned in life and how people can learn from their experiences to improve lives instead of repeating their mistakes that many people still ignore at their own peril.

To book Dr. Nathanael-Israel Israel for a speaking engagement purpose, visit Science180.com/speaking.

How you can make money by joining the affiliate program to sell Nathanael-Israel Israel's books

Greetings,

Do you want to make easy money by selling the #1 universe-origin, life-origin, chemical-origin, and scientific proofs of God's existence books on your website, newsletter, network, and by mail? You can start making money as you help sell Science180 Books, including this one, which specializes specialized in helping people across the globe to scientifically decode and understand the formation of the universe, life, and chemicals.

Your contacts, site, blog, forum, podcast, and newsletter may be admired among my target audience. Some of my products and services may interest your audience. My books are the first in history to scientifically demonstrate the match between science and Biblical creation in a way that satisfies both believers and nonbelievers, a historic achievement and discovery that is revolutionizing our view of the origin of the universe, life, and chemicals for the benefit of humankind.

Imagine you have a website where you can talk to people about my books and services and get a great percentage of every purchase they make on my site. Imagine you send a certain link about my books to your friends or network, and when any of your contacts buy a copy of my books, you get a percentage or a certain amount of what they pay on my sites. Imagine you can email your friends and spread the good news about my books, and when anyone uses that link to buy my books, I give you something. Well! This is what the affiliate program is about. Apply today or learn more about it at Science180.com/affiliate. Likewise, if you own a website, you can apply for Science180's affiliate program and get a specific affiliate link that you will place on your website and newsletter, and if people click on it to buy my books, they will be led to my page, and after they buy, I will pay you a certain amount, sharing the profit with you instead of just verbally saying thank you.

Would you be interested in reviewing some of my products and services with the aim of becoming an affiliate? We have a wonderful affiliate program, and commissions are paid quickly and accurately.

If you are satisfied by the quality of our products and services, I am convinced you will also be impressed by our affiliate program.
I look forward to hearing from you

Nathanael-Israel Israel, PhD

Collaborate or partner with Nathanael-Israel Israel

If you have any lawful idea, initiative, or suggestion for a genuine partnership with Nathanael-Israel Israel or Science180, please visit Science180.com/partner to inform us.

How to be trained or mentored by or have a one-on-one consultation with Dr. Nathanael-Israel Israel

Hire Nathanael-Israel Israel to train you or your organization in the best ways to conduct yourself and your organization to align your initiatives with the real understanding of the origin of the universe, of life, and of chemical particles in a way that you will not hear anywhere else. Nathanael-Israel Israel offers training through the program called "Science180 Academy." For training purposes, please visit www.Science180Academy.com.

Visit Nathanael-Israel Israel's personal website to get great resources you won't find anywhere else for free.

To stay in touch with Nathanael-Israel Israel, and to get updates directly from him, please visit his website and sign up for his popular newsletter at Israel120.com/newsletter.

Ask for review

If you are a book reviewer or a professional wanting to review this book or others written by Nathanael-Israel Israel, please contact us at Science180.com/AskForReview

Donate and support Nathanael-Israel Israel's efforts and initiatives

To help humankind accurately understand the real origin of the universe and its content, like I have done in the groundbreaking books I published after 12 years of sacrifice, I need your financial support. Please consider donating to me or to Science180 by visiting Israel120.com/donate or Science180.com/donate.

Your donation will be used to help me continue doing what I did to birth these books that you enjoyed and that you know will help many people across the globe. No amount of money is too small or too big. Whatever you can give, please give.

Quantity discounts: Purchase Science180 books, including this one, in bulk at a special discount

To purchase Science180 books, including this one, in bulk at a special discount for sales promotion, corporate gifts, fundraising, or educational purposes, or to create special editions to specifications, contact specialsales@science180.com or visit Science180.com/discount.

Buy a copy of Nathanael-Israel Israel's books for your friends, family, or anyone

If this book has been a blessing to you, and I know it has, please consider getting another copy and giving it to a friend, a family member, or someone you think it may help or challenge. If you want to get many copies, we can even give you a discount;

just contact us as we previously explained.

Recommend Nathanael-Israel Israel's books to your organization

Because I know this book has been a blessing to you, I ask that you recommend it and others that I wrote to your organization, class, workplace, church, school, network, or clubs. Recommending this book will help others to tap into the blessing and opportunities that my books will open for them.

Share Nathanael-Israel Israel's groundbreaking discovery with others

To improve more lives, please share the findings of Nathanael-Israel Israel's books with others, for many people out there still do not understand how the universe was formed, and sharing your experience of reading this book will help them. If you enjoy Nathanael-Israel Israel's books, please help other people find them by writing a book review on your blog or on online bookstores, or write it and share it with us. Likewise, share and mention this book on your social media platforms (e.g., Facebook, Twitter, YouTube, etc.).

Follow Nathanael-Israel Israel on social media

In our modern world, social media has become a huge part of how messages spread across the globe today. To ensure more people hear about the good news revealed in my books, I need you to follow me and share my content on your social media and in your network. To know the full list of my social media accounts and follow me, please visit Science180.com/socialmedia.

Share your feedback, criticism, testimony, experience, adventures, story, or comment about this book with me

How have Nathanael-Israel Israel's books and services at Science180 improved your life? I would love to hear from you.

To help me know how I can better help you next and encourage others, I need to know and capture your testimony or criticism. Please visit the feedback page, Science180.com/feedback, to tell me:

- how this book impacted you or will impact you
- what you like or dislike or disagree with
- what you think, wish, or dream that I need to work on next
- what you wish to see in this book but that was absent

- what shocked you the most
- what got your heart pumping as you were reading this book
- what you found more insightful or thought-provoking
- what you want to do to be a part of my journey
- how my work changed your life or someone else's life

Message from the publisher of this book

Just like Nathanael-Israel Israel, you can publish your book(s) with us too. To get started and see how we may help you, please visit Science180Publishing.com today.

To contact Nathanael-Israel Israel or Science180

For any suggestions or questions, please visit Science180.com/contact and Nathanael-Israel Israel's personal website: Israel120.com. Feel free to ask me any questions you have about the universe's formation, life formation, and chemical formation.

12. 'SCIENCE180 ACADEMY' SUCCESS STRATEGY

Science180 Academy

Science180 Academy is a training, speaking, consulting, and mentoring program designed to groom and empower people of all backgrounds in the truth about the origin of the universe, life, and chemicals. According to their background and interest, trainees are taught different levels of scientific facts to grasp a deeper understanding of the origin of the universe and, unfortunately, less analyzed. If you want to be enlightened and equipped so you can cause positive changes in your respective field of expertise, then the Science180 Academy program is for you.

Science180 Academy does not confer college credit, grant degrees, or grade its attendants, participants, or students. It is not an accredited university or college but is the one-stop destination for universe-origin, life-origin, and chemical-origin experts. It is where scientists and laypeople get all their origin-related questions properly answered. It is the only place where the accurate interpretation of the universe-origin, life-origin, and chemicals-origin data matters a lot.

Science180 Academy brings together Dr. Nathanael-Israel Israel (the founder of Science180) and other experts to deliver outstanding value, insight, and lessons to assist you to accurately understand the true origin of the universe, chemicals, and life, so you can tap into that knowledge to improve lives perpetually. Nathanael-Israel's goal is to give you practicable and undeniable proofs of the formation of the universe so you can be fired up to become the best version of yourself and to cause positive changes to your initiatives that will profit you today and forever. For Nathanael-Israel, decoding the origin of the universe and everything in it is not a job but his life mission, and helping others to fully understand that is his mission. Visit Science180Academy.com today to start.

If you are still wondering if Science180 Academy is for you, let me also inform you that some of Science180's clients and prospects have a profound technical knowledge and background in science, while others don't. Some are creationists (e.g., Science180 creationists, Young Earth creationists, Old Earth creationists, and Intelligent Design proponents); others are anti-creationists. Some are believers; others are freethinkers (including atheists, humanists, rationalists, agnostics, nontheists, nonreligious, skeptics, nonbelievers, religiously unaffiliated, spiritual-but-not-religious, ex-believers, and doubters). Regardless of their background, belief, or disbelief, Science180 works with each of these people to figure out their needs, priorities, and the products and services that best fit them. Science180 improves their knowledge, experience, and performance and answers their questions (related to the universe's origin, life's origin, and chemicals' origin) by crafting a personalized program that perfectly matches their interests, needs, and things that are dear and meaningful to them, whether it is to:

- Protect yourself and loved ones by keeping all of you secured and empowered with the true knowledge of the origin of the universe
- Have a reliable access to the world's authority on origin-related matters and get your origin questions professionally answered with the truth step-by-step
- Connect with practical tips about how to decode the origin of the universe, life, and chemicals and protect yourself from wrong theories in the literature and the media
- Get inside secrets about how to locate flaws in origin-related theories so you can save time, money, and other resources to improve lives
- Ultimately boost your confidence in detecting, confronting, and avoiding wrong theories by knowing the facts and processes involved in the formation of the universe
- Enjoy multiple origin-related programs and choose the ones that best suit your needs
- Bypass technical knowledge that restricts non-experts from accessing the origin-related truth contained in the massive scientific data, and get to the bottom of scientifically-locked origin-related secrets regardless of your background
- Benefit from continual updates and assistance during your journey to decode the universe, and clear your way for the universe-origin related freedom, power, technology, innovation, and breakthroughs of the future
- If you are a teacher, discover tools to better engage with students and introduce them to new ideas
- If you are a physicist, understand the mother of all turbulences that shaped everything in the universe
- Scientifically test and know whether there is a God that created the universe or not, and which God it is
- Free yourself from boring explanations of the origin of the universe, life, and chemicals and embrace the proven theory that opens doors to unparallel opportunities
- Disrupt all religious and scientific chains of repetitive nonsenses about the universe's origin and life's origin, and turn your attention toward unconventional ideas leading to greater innovation and prosperity

- Satisfy your burning desire for freedom from beliefs and scientific theories about the universe's origin and life's origin that suffocate you and bind your mind, faith, unbelief, heart, and education
- Empower yourself to leave unforgettable marks and to stand tall as a symbol of freedom, power, creativity, and originality in your field of expertise
- Fearlessly push the boundaries of the human abilities to properly understand what is perceived as un-understandable, mysterious, supernatural, unimaginable, impossible, and unthinkable that holds you back
- Make a difference and blaze new trails for those who depend on your leadership
- Empower and align yourself with Science180, the historic company that has done what no other organization has ever done: accurately decode the origin of the cosmos and its content

To register or to learn more, visit Science180Academy.com today.

Now, I will present some Science180 Academy programs. Owned by Science180, Science180 Academy specializes in everything at the intersection of reason and faith, or science and religion.

Science180 Academy deals with different subjects according to the needs of its members or target groups. When people register for Science180 Academy, they must choose the program(s) they want to focus on so their training can be properly personalized accordingly. This is similar to how people register at a university and take classes in a specific department matching their needs!

Science180's breakthroughs are so complex and dense that it is not realistic or good to try to explain them all in just one academy, else people will be overwhelmed, disinterested, and confused by the plethora of data to handle. In other words, Science180 Academy offers a wide range of origin-related training in various domains strategically designed to allow people to choose the most suitable for their needs so that, regardless of their background or field of expertise, people can equip themselves, align their mindset, and improve lives today and forever using the accurate explanation of the origin of the universe, of life, and of chemicals. The Science180 Academy curriculum is based on 12 years of deep, unconventional research that culminated with the publication of many much-admired books on the formation of the universe and its content. The content of each Science180 Academy is strategically crafted by Dr. Nathanael-Israel Israel (who is acknowledged as the internationally acclaimed world's authority in origin-related issues) to suit both scientists and nonscientists, religious and nonreligious people, and leaders as well as followers, so they can fully decode the proofs of the formation of the universe, of life, and of chemicals they have been wanting to demonstrate or grasp.

The current programs of Science180 Academy are

- **1. SCIENCE180 ACADEMY OF COSMOLOGY** (Designed for all scientists who want to scientifically study cosmology, the science of the origin and fate of the universe)

- **2. SCIENCE180 ACADEMY OF TURBULENCE** (This is a perfect fit for scientists and other experts interested in studying abiotic turbulence). Examples of these people include:
    - Automobile manufacturers
    - Chemical producers
    - Commercial space businesses / aerospace manufacturers
    - Government space agencies
    - Private aerospace companies / aircraft manufacturers
    - Technology companies
    - Turbulence companies, experts, and scientists

- **3. SCIENCE180 ACADEMY OF LIFE SCIENCES** (Tailored to those who want to study biotic turbulence):
    - Life science specialists, associations, or organizations
    - Healthcare companies (pharmaceutical, medical, health insurance and technologies)
    - Biotechnological companies

- **4. SCIENCE180 ACADEMY OF CHEMISTRY** (Designed for chemists, biochemists, scientists, and other educated people who want to understand the origin of chemical particles)

- **5. SCIENCE180 ACADEMY FOR LAYPEOPLE OR THE GENERAL PUBLIC** (Very fit for any layperson or "less" educated people who want to learn (in a simple language) deep insights that even those who went to university for years were unable to decrypt by themselves, so these laypeople can be equipped to eliminate all forms of scientific and religious universe-origin prejudices.)

- **6. SCIENCE180 ACADEMY FOR CHILDREN** (This academy breaks down origin key topics into language that children can fully understand). This is the only Science180 Academy that your whole family will like and enjoy together, and which will set children on the path of success by accurately showing them early in life the formation of the universe, and that will teach them how to detect errors in theories or stories that would misguide them as they grow up. Therefore, you need to add this great, efficient, trustworthy, and cost-effective "Science180 Academy for Children" to the strategic journey of children toward their best tomorrow.

- **7. <u>SCIENCE180 ACADEMY OF THE PSEUDEPIGRAPHA AND SPIRITUAL WORLD</u>** (Only one ancient blueprint has the reliable power to help you to accurately decrypt the spiritual origin and history of everything in the universe. If you are a believer and want to delve into the prophetic, angelic, and higher order of knowledge based on the spiritual world, then this Science180 Academy is for you. This program is suitable for those who took at least "Science180 Academy of Creationism". For you to enjoy the courses in this Academy, you need to have at least learned about or attended Science180 Academy of Creationism. If not, you may waste your time trying to grasp simple and supernatural things that cannot be scientifically proven in this academy, but in Science180 Academy of Creationism).

- **8. <u>SCIENCE180 ACADEMY OF CREATIONISM</u>** (Science180 Creationism is a scientific theory spearheaded by the groundbreaking discoveries of Nathanael-Israel Israel, that scientifically explained the origin of the universe, life, and chemicals using turbulence and that mathematically reconciled science and the Biblical account of creation for the first time in history. Science180 is different from all existing creationist theories known before 2025. Science180 Creationism reconciled science with the Biblical account of creation, including scientifically proving that the Earth was formed on Day 3, while the Moon and the Sun were formed on Day 4 of creation). As you attend "Science180 Academy of Creationism", you will receive accurate answers to all your questions concerning the creation of the universe. The target audience of "Science180 Academy of Creationism" includes:
  - o   Christians of all ages and all educational backgrounds
  - o   Christian or Bible colleges, universities, and schools
  - o   School boards of education and textbook initiative leaders
  - o   Christian ministries and associations
  - o   Creationism organizations
  - o   Anti-creationists wanting to explore the Biblical creation narrative

- **9. <u>SCIENCE180 ACADEMY FOR FREETHINKERS & ALL ANTI-CREATIONISTS</u>** (This Science180 Academy is designed for evolutionists, anti-creationists, and all other types of unbelievers seeking to rationally explore and understand alternative arguments for the creation or formation or origin of the universe, life, and chemicals from a fresh, scientific perspective).

- **10. SCIENCE180 ACADEMY OF LEADERSHIP**-(Also called "Science180 Academy for Leaders", this program will enlighten leaders of organizations on how to solve their people problems, process problems, and profit problems related to the origin of the universe, of life, and of chemicals according to their domain of expertise). With "Science180 Academy of Leadership", leaders will gain new insights so they can cast new visions and avoid focusing on screwed-up processes, products, and services related to universe-origin initiatives that need to be fixed, faced, or dealt with. Science180 Academy of Leadership will also equip leaders to address problems related to inefficiency, gaps, missed opportunities, wasted time and efforts, too many steps, bureaucracy, useless layers between organization and customers concerning the innovation, research methodology, research, product development, strategic planning, workforce diversity in alignment with the historic Science180 breakthroughs so that they can sell more often at full price, avoid regrets in the end, open new markets focusing on real solutions, expand their products and service lines, cut useless costs and research, stop wasting time on useless products that will yield nothing, start focusing on the real money-making problems, reach and convert more prospects into profitable and loyal customers, speed up time to market, avoid spending resources on unprofitable projects but on profitable ones, take their organization to a higher level, open new groundbreaking doors, boost their margins of profits, beat their competitors, and make big profits, conserve more cash and spend it on more valuable things, and outpace their competition as long as their products or services are related to the origin of the universe, life, and chemicals. Perfect fits for Science180 Academy of Leadership include leaders of:
  - o Scientific organizations
  - o Academies of sciences
  - o Universities, colleges, and schools
  - o Automobile manufacturers
  - o Government space agencies
  - o Commercial space businesses / aerospace manufacturers
  - o Private aerospace companies / aircraft manufacturers
  - o Technology companies
  - o Healthcare and biotechnology companies
  - o Any other organization that can benefit from the insight into the origin of the universe, life, and chemicals

- **11. SCIENCE180 ACADEMY FOR GOVERNMENTAL AGENCIES** (Do you want to know how and why most nations and governments are wasting millions of dollars on universe-origin and life-origin research they don't need ... and how to avoid it? Indeed, for most developed nations, and even for some underdeveloped countries, universe-origin projects can cost billions of US dollars and other expensive things that cannot be afforded without sacrificing crucial priorities. Even in developed countries, the impact and the return on investment of the space researches are the subject of intense political and economic debates. What if your nation or institution could reduce wasteful spending on universe-origin research and life-origin research, as well as your dependency on wrong theories on the origin of the universe and life? "Science180 Academy for governmental agencies" will show you how to use the latest scientific breakthrough to better understand the origin of the universe without wasting money on what is already known or what we think we don't know but that most scientists ignore. Having spent years accurately decoding the origin of the universe, of life, and of chemicals, Dr. Nathanael-Israel Israel delivers science-backed insight to properly understand all the processes connected to the universe's formation—so innovative. Perfect matches for "Science180 Academy for governmental agencies" include leaders of:

- o Department of Defense including its military branches
- o Department of Energy
- o Aeronautics / Space agencies
- o National Science Foundation
- o Other key governmental agencies

12. <u>OTHER SCIENCE180 ACADEMY:</u> If you did not relate with any of the Science180 Academies (see Science180Academy.com) mentioned above, but you are still interested in learning something specific about the origin of the universe, life, and chemicals that better fits your needs, please visit Israel120.com to contact me so we can discuss that with you.

Science180 Services and Products you will love

Because you are reading this book, you are probably very interested in answering your questions about the origin of the universe, of life, and of chemicals. Imagine you want to be trained by Dr. Nathanael-Israel Israel and his team so you can benefit from their outstanding expertise to empower yourself or your team. Or you want him to give a keynote speech, a seminar, or any other kind of talk or conference at your organization. Or you want him to mentor you or some people or a team at your organization. Maybe you have critical origin-related questions that you need his help to accurately answer. You want a true expert to talk with you about the customized program or game plan that fits your needs. You want him to tailor his advice, expert feedback, and proven shortcuts to the stage of life you are in and help you get to where you want to be in your desire to properly understand the origin of the universe, life, and chemicals and harness the benefits that come with it. Perhaps you don't know how to properly get any of these important tasks done according to your specific needs or the needs and demands of your organization. That is what Science180 Academy is all about. Visit Science180.com/services for more details about how to benefit from the services that Science180 provides.

Maybe you are a leader that wants to hire Dr. Nathanael-Israel Israel and his team to train some departments at your organization. Or you want to refer them to other companies like a good dish passed around the dinner table, and you want to explore how Nathanael-Israel Israel can pay you something for that referral. Maybe you attended Nathanael-Israel Israel's speaking program, for which, without going into details, he accurately raised your awareness about how the universe, life, and chemicals were formed. Or maybe you attended his training, in which he detailed and showed you how he decoded the scientific data using various tools and certain thinking strategies that helped him and which transferred some skills to you; and now, you are interested in a long-term one-on-one consulting or mentoring program with him so that he delves into more details about how to use proven techniques to decode the universe (strategies for data collection, data analysis, data presentation, writing, and even tips for future research) and change your behavior on a long -term basis. If you relate to any of the points mentioned above, Science180 Academy is the right fit for you!

Other customizable services that Science180 provides include:

- Assessments
- Book publishing (Yes! Science180 can publish your books!)
- Conferences
- Consulting
- Executive mastermind groups
- Face-to-face visits
- Master classes
- Online courses
- Podcasting
- Seminars
- Speaking engagements (offline and onsite—e.g. seminar, keynotes, and workshops)
- Survey and research tools
- Training
- Video programs
- Virtual presentations

Here are other reasons why you should choose to work with or hire Nathanael-Israel Israel and the team at Science180:

- A simple God's existence theory that made no assumption
- All the products and services you need to accurately and easily decode

God's existence
- Breaking traditional nonsenses about God's existence
- Bringing people together through the power of the accurate decoding and understanding of science and God's existence
- Complex God's existence questions solved accurately in a simple language
- Discover the key variables needed to decipher God's existence
- Easily understand complex God's existence equations in minutes
- First stop for your God's existence needs
- God's existence formula accurately made easy
- God's existence inerrancy guaranteed
- God's existence theory that helps you fight wasteful programs
- Historic and accurate God's existence breakthrough
- Improve your understanding of God's existence with new, accurate products and services
- Mover of the needle on science vs. creationism debate
- Nonconformist, rule-breaker, and accurate demonstrator of God's existence
- One-stop for answering God's existence questions
- Personalized God's existence decoding package
- Reinvent our rational interpretation of the Genesis story of creation
- Reuniting science and God's existence
- Science and God's existence reconciliation made possible
- Science180: The premier organization that scientifically decoded God's existence accurately
- State-of-the-art decoding experience of the existence of God
- The all-in-one proven & uncomplicated God's existence formula
- The formula at the intersection of science and God's existence
- The go-to source for valuable God's existence information
- The new physics that will revolutionize God's existence debates forever
- The place where scientists and laypeople get all their God's existence questions properly answered
- The science that refreshes your mind and faith or doubt
- Undeniable reconciliation of science and God's existence
- Understand God's existence. Increase your glory and peace of mind

Learn more at Science180.com/services.

Science180 Models of the origin of the universe and its content

Science180 Models consist of all the theories elaborated by Nathanael-Israel Israel regarding his groundbreaking discovery on the origin of the universe and its content, including all forms of life and chemical particles. These theories are detailed in various books written by Dr. Nathanael-Israel Israel and encompass the following:

- **1. *SCIENCE180 MODEL OF COSMOLOGY***
- **2. *SCIENCE180 CREATIONISM***
- **3. *SCIENCE180 MODEL OF THE ORIGIN OF CHEMICAL PARTICLES***
- **4. *SCIENCE180 MODEL FOR THE GENERAL PUBLIC***
- **5. *SCIENCE180 MODEL OF LIFE-ORIGIN***
- **6. *SCIENCE180 MODEL FOR CHILDREN***
- **7. *SCIENCE180 MODEL OF PSEUDEPIGRAPHA***
- **8. *SCIENCE180 MODEL OF THE PROOF OF THE EXISTENCE OF GOD***
- **9. SCIENCE180 THEORY OF EVERYTHING**

Learn about them at www.Science180.com/Models

'Science180 Academy' Success Strategy
SCIENCE180 PUBLISHING: AUTHORS WANTED

Science180 Publishing, the American publishing company that published the groundbreaking discovery about the origin of the universe, of life, and of chemicals spearheaded by Dr. Nathanael-Israel Israel, really wants to publish your book(s) regardless of your field of expertise. This is a unique opportunity for:

- established authors
- people aspiring to become authors
- people who have written a book or are wanting to write one and need help with anything regarding publishing
- people who are not well known, inexperienced
- people whose books are viewed as nonconformist, controversial, or unconventional
- people who do not have enough resources or knowledge to navigate the publishing process
- people who are struggling to find an affordable, experienced, and high-quality publisher

Although Science180 Publishing is based in the USA, it can publish your books within your budget regardless of your geographical location. Science180 Publishing is highly interested in your document and possibly helping you publish it. Please visit Science180Publishing.com to explore how we may assist you. No matter the content of your book, as far as it is original, not promoting anything illegal, not duplicating anyone else's idea, Science180 Publishing can help you publish it in the USA. Please contact us asap and see how we can help.

To start your journey of publishing your book with Science180 Publishing, please visit Science180Publishing.com today.

Science180 Interview Report (aka Science180 Internet-TV-Radio Interview Report)

Science180 Interview Report is the newsletter to read for guests and unconventional show ideas at the intersection of science and faith. Indeed, many hot questions are still unanswered on the road leading to the correct understanding of the origin of the universe, of life, and of chemicals. But most people don't know where to find the accurate answers to those challenging questions. What if with one simple call you could accurately answer all of those questions? You need to get in touch with or interview Dr. Nathanael-Israel Israel on your show, radio, TV, podcast, and even website, or invite him for a live presentation at your organization if your audience can benefit from any of the following show, talk, speaking, or interview ideas:

- Are most Christians denying the God they want the nonbelievers to accept?
- Can anyone really be scientifically 100% sure and prove that God created the universe?
- Can Christians fix the growing trend according to which more people are denying God in favor of secular theories because they think that it is impossible for science and faith to meet?
- Can atheists, rationalists, and all other freethinkers use science to talk themselves out of denying God?
- Can mathematics and science rescue Christians in their efforts to rationally prove the existence of God, the Creator?
- Can most Christian leaders refuse to take a stand on 6 literal days of creation and expect atheists and freethinkers not to argue that God is simply unnecessary?
- Can we explain the formation of the universe through natural processes without evoking evolution and the Big Bang?
- Can we mathematically prove that the formation of the Earth was completed on the 3rd day of creation, like the Bible says?
- Can we scientifically demonstrate without a doubt that the Moon and the Sun were really formed on the 4th day of creation, like the Bible says?
- Did the Quran or any other religious book make any gigantic error about the universe's creation that any scientific formula proves the Bible got it right— and vice versa?
- Does the Bible scientifically teach anything about the universe's origin that most people, including Christians, ignore?
- Does the Bible scientifically teach anything about the origin of life that most people, including Christians, ignore?
- How can people approach difficult topics such as people challenging the accuracy of the Biblical account of creation, the Big Bang theory, or any other theory on the origin of the universe?
- How to scientifically talk about the universe's formation and have nonbelievers and all other freethinkers rationally agree?
- Is it a waste of time to attempt to prove the Biblical creation using science or historical investigation?
- Is science making you doubt God or the Bible?
- Is science making you doubt your faith?
- Is the Bible an obstacle to scientific progress?
- Is the Bible's account of creation making you doubt science?
- Is there any need to prove the Biblical creation to be true?

- Must Christians apologize to atheists, rationalists, and all freethinkers for the proofs creationist scholars and preachers have used to demonstrate creation?
- Why are most nations wasting millions of dollars on universe-origin and life-origin research they don't need—or do they?
- Why do Christians hate theories that compromise with Darwinism and the Big Bang?
- Why doesn't the secular world care much if Christians and their leaders believe in evolutionism, but they actually care much if they don't believe in the billions of years process?

I know you may be tempted to answer these questions by yourselves, but avoid landing yourself on wrong paths that caused some people to lose contact with reality. It is better to get the accurate answer from the know-how expert, Dr. Nathanael-Israel Israel, the author of many books on the origin of the universe, of life, and of chemicals, and the standout expert who accurately decoded the scientific formula that forces science to bow to the truth. If you would like to register for the Science180 Interview Report so we can periodically send you show ideas and opportunities related to the origin of the universe, of life, and of chemical particles, please visit Science180Interviews.com for more details. Invite Nathanael-Israel Israel to your organization to discover his unique answers to these critical issues.

Back cover of other books by Nathanael-Israel Israel
TURBULENT ORIGIN OF THE UNIVERSE
THE FIRST AND ONLY SCIENTIFIC BOOK THAT ACCURATELY EXPLAINS EVERYTHING YOU NEED TO UNCONVENTIONALLY, EASILY, AFFORDABLY, AND ENJOYABLY DECODE THE UNIVERSE FORMATION
In *"Turbulent Origin of the* Universe," filled with great diagrams and digestible scientific facts, you will discover, learn, or get:

- The all-in-one, proven & uncomplicated scientific formula that accurately decoded the formation of the universe, and that explained the birthdate of the stars, planets, satellites, asteroids, and all other celestial bodies in the universe, so you can position yourself to stay on top of your competitors, avoid repeating crucial mistakes that many people have ignorantly made at their own perils
- Extraordinary, unprecedented, accurate insights into the first factors (e.g. early universe physics) that defined the history and formation of the universe so you can tap into deep scientific secrets you ignore, and set yourself apart from others
- The new physics that will revolutionize science forever and land you into a

zone of original ideas that improve lives nonstop regardless of your expertise

- The 4 simple things without which it is impossible for anyone to ever understand the formation of the universe, think accurately, work differently, achieve, or perform better for superior results

- The verified key to move the cosmological mountains of misunderstanding, so you can confidently free your mind from doubts, improve your health, and prevent you from any danger connected with sticking with wrong assumptions

- Save time and money, and enjoy your life once you remove errors holding your true understanding of the universe's origin captive

- Historic scientific proof of whether a planet was formed in 2.82 days, whether a satellite was formed in 3.32 days, and whether a star was formed in 3.69 days after the beginning of the universe, so you can creatively produce and address a broader work spectrum by learning how to effectively communicate with and establish unusual connections between otherwise disconnected and disparate scientific data

- The scientific formula that successfully tested the existence of God in a way that shocked believers, skeptics, and all other freethinkers

- Why the scientific community has failed to sufficiently explain the origin of the universe and understand how existing theories have missed and undefined central ideas, and imposed limits on the vision of scientists

- Specific in-depth knowledge, up-to-the-minute information, and ideas so you can expand your market, cut useless costs, stop wasting time on inadequate projects, and start focusing on the profitable solutions (Science180.com/scientific)

- How Science180 Academy can strategically enlighten you, guide you to navigate and filter the massive data collected on the universe, so you can answer the world's most challenging questions, remove any scientific and philosophical cataracts that may be blocking you, and bring you many steps closer to your best life

- How to better resonate with your target market that is craving something original that breaks wrong explanations of the universe's origin

Get *"Turbulent Origin of the Universe"* today to begin an incredible journey of accurately decoding the universe and change your life forever!

Dr. Nathanael-Israel Israel is told by people that he is the #1 universe-origin, life-origin, Life-origin, and chemical-origin expert. He is the founder of Science180 and the author of many books on the origin of the universe and its content. To learn more about how he may help you, visit Israel120.com.

RECONCILING SCIENCE AND CREATION ACCURATELY

THERE IS ONLY ONE SIMPLE, COMPELLING, SOLUTION-DIRECTED SCIENTIFIC FORMULA ACCURATE ENOUGH TO RATIONALLY EXPLAIN HOW GOD CREATED THE UNIVERSE

"Reconciling Science and Creation Accurately" is a landmark book in universe-origin writing from a rare perspective by one of the most respected minds of our time. It scientifically explores the most challenging questions of all time that believers, nonbelievers, and all freethinkers are interested in: How can we rationally demonstrate, without checking our brain at the door in the name of faith, that God created the universe? How did the universe begin, and what processes did God use to create it? Are these processes still operating in the universe or not? Can believers abandon wrong theories if they think it is impossible for science to literally prove the Genesis story, or if they think that science is evil and diametrically opposed to faith, or if they compromisingly embrace scientific theories that contradict the Biblical account of creation written before the scientific era? What can believers do to help the skeptics believe in the biblical narrative of creation?

Lucky you, Dr. Nathanael-Israel Israel successfully navigated those questions with an accuracy that both scientists and nonscientists have been applauding across the globe. After reading *"Reconciling Science and Creation Accurately,"* you will confidently:

- Scientifically prove the Biblical account of the creation of the universe and the existence of God in a way that makes the head of those who deny God spin faster than a DJ's turntable
- Know how to rationally talk to anti-creationists, evolutionists, Big Bang proponents, atheists, skeptics, and other freethinkers about the universe's formation, and they will beg you to know more about God, the Creator, that they mistakenly rejected
- Discover very accurate, rare, and factual truths about the universe's origin that will save you time and money, and get you much closer to the better and joyful life you want to live today and forever
- Improve your health and faith by knowing that the existence of God can be scientifically justified using Science180 Cosmology and particularly Science180 Creationism
- Enter a new area of freedom and power by crushing the head of and breaking free from the suffocating expectations of all wrong theories that have hijacked secular and religious education, and that have held the Biblical account of creation captive for almost 3500 years
- Break free from the suffocating expectations of some forms of creationism that have sequestered the mind of some believers for a long time
- Uncompromisingly, intelligently, and scientifically explode the myth of those who, instead of literally taking the Biblical days of creation as 24-consecutive hour days, think that they were millions of years, or were representative of

long ages, or that millions of years existed before them or were positioned between them

- Understand the accurate standard to interpret the Biblical account of creation thanks to Science180's breakthrough that transformed science and laid a foundational bedrock for the inerrancy of Scripture

Now that Genesis (the oldest manuscript in the world, written before science and most religions were born) is scientifically proven to be correct (Science180.com/biblical), what unstoppable, jaw-dropping paradigm shift will the discovery of the perfect alignment between science and the Bible bring into the religious, rational, and secular world today? Get this thoughtful book now to figure out what happened at the beginning, what is coming up, and why it is time to urgently rethink everything you have been told about the universe's origin so you don't eventually regret it! Don't say nobody told you!

Founder of Science180 Academy, **Dr. Nathanael-Israel Israel** is acknowledged worldwide as the discoverer of the all-in-one, proven, and simple scientific formula that accurately cracked the origin of the universe, of life, and of chemicals, and that scientifically unearthed the holy grail at the intersection of science and the Biblical account of creation. Learn more at Israel120.com.

TURBULENT ORIGIN OF CHEMICAL PARTICLES

FIND ALL THE RELIABLE, CONVINCING, SCIENTIFIC ANSWERS YOU NEED TO SUCCESSFULLY DECODE THE ORIGIN OF CHEMICAL PARTICLES SAFELY

Where did all elementary particles and composite particles, including atoms, molecules, minerals, and rocks, come from? What are the fundamental factors, the machinery, and the generic processes that defined their formation and proprieties? What was the nature of their precursors at the beginning of the universe, and what underlying processes shaped or molded them into the chemicals we know today? What was the primary cause of the abundance and diversity of chemicals in the celestial bodies in the universe? What is the accurate link between the formation of chemical particles and the formation of galaxies, stars, planets, asteroids, and satellites? What light can the origin of chemicals shed on the real cause and meaning of gravity and the other so-called fundamental forces in nature? How does the formation of the chemical particles fit into the big picture of the formation of the universe?

After studying these questions for more than 12 years, Dr. Nathanael-Israel Israel discovered that the proper understanding of the origin of chemical particles is a very challenging but profitable task that requires original, scientific, mathematic, and philosophic efforts beyond the current state of modern science—until recently. The solution for all of these puzzling problems is problems: *"Turbulent Origin of Chemical Particles,"* the straightforward and trustworthy book that will help you to quickly, cheaply, easily, and efficiently navigate everything you need to know to finally solve the hard problems about the origin, the formation, and the functioning of all chemical particles. Whether you are a chemist, a biochemist, or any other scientist or engineer, as long as you have a reasonable background in chemistry but ignore how to scientifically demonstrate the origin of all chemical particles, this marvelous book is for you!

Amazingly packed with eye-popping analysis, fantastic graphs, tables, and the historic formula that broke the universe-origin code, *"Turbulent Origin of Chemical Particles"* will:

- Make it easier than ever for you to properly understand, decrypt, and articulate the real origin of natural chemical particles in the universe, therefore freeing you from false and boring explanations of the origin of all matters, and embrace the proven theory that opens doors to unparallel opportunities
- Professionally teach you how to transform the true knowledge of the origin of chemical particles into insights that significantly add value to your life in less time, and successfully establish you as a symbol of freedom, power, creativity, and originality in your field of expertise
- Fire you up to become the best version of you, and to cause positive changes to your initiatives that will profit you nonstop
- Discover thrilling illustrations and unconventional explanations of the formation of all matter in the universe, written in a simple language that brings humankind much closer to the complete deciphering of the mysteries at the very heart of chemistry, and open the way to a future of technology, innovation, discoveries, and breakthroughs
- Equip you to bypass technical knowledge that restricts non-experts from accessing the origin-related secrets contained in the massive scientific data, and get to the bottom of origin-related mysteries regardless of your background so you can empower yourself to leave unforgettable marks in your field of expertise
- Learn more at Science180.com/chemical

With *"Turbulent Origin of Chemical Particles,"* the accurate decrypting and understanding of the formation of chemicals has never been profitable and easy. Hence, this great book is THE ultimate how-to guide for great people wanting to correctly decode the origin of the chemicals and positively transform their lives. Get this celebrated book today. Don't wait!

Known as the nonconformist, rule-breaker, and accurate demonstrator of the universe's origin, **Dr. Nathanael-Israel Israel** is the founder of Science180, the one-stop for answering the most crucial universe and life's origin questions. He has had the honor of being acknowledged as the fearless universe-origin decryption trailblazer. Learn more at Israel120.com.

ORIGIN OF THE SPIRITUAL WORLD
ONLY ONE ANCIENT BLUEPRINT HAS THE RELIABLE POWER TO HELP YOU TO ACCURATELY DECRYPT THE SPIRITUAL ORIGIN AND HISTORY OF EVERYTHING IN THE UNIVERSE

Countless books talk about the origin of the universe and of life, but this amazing book is the first and the only one that has undeniably explained how the formation of the universe and everything in it was truly revealed in the rejected and hidden scriptures such as the Books of Enoch and others. In *"Origin of the Spiritual World,"* you will:

- Discover deep rejected secrets that have prevented humankind from unearthing the beginning of the universe
- Plainly see the scientific proof (hidden in scriptures) of the formation of the Earth, the Moon, and the Sun in a matter of days, a historic revelation that bizarrely and shockingly matches the scientific data as scientifically proved in *"From Science to Bible's Conclusions"*, a popular book written by Dr. Nathanael-Israel Israel
- Properly use the lost and rejected scriptures to articulate the process by which the universe was formed, and use that insight to improve your understanding of the Bible, innovate in your domain of interest, and improve your life perpetually
- Empower and align yourself with the historic breakthrough that has done what no other discovery has ever done: accurately unlock and decode mysteries concerning the origin of the cosmos and its content using scientific keys revealed in ancient scriptures that some elites have concealed (Science180.com/pseudepigraphic)
- Discover and apprehend the complex formation of the universe and life without leaving out the challenging questions that people of all ages have been struggling to answer for thousands of years, while the answers were hidden
- Find more joy in life through a clear interpretation of old and fresh revelations about the creation of the universe astonishingly backed by modern science, which some people wrongly think opposes the Bible
- Make a difference and blaze new trails for those who depend on your leadership

If you believe in God, have some origin-related questions whose answers you cannot find anywhere, not even in the Bible, and want to tap into historically neglected revelations to answer fundamental universe and life questions, then be sure to get a copy of *"Origin of the Spiritual World"* today.

Dr. Nathanael-Israel Israel happens to be the discoverer of the historic mathematical equations that scientifically demonstrated that the Earth was formed

2.82 days, the Moon 3.32 days, and the Sun 3.69 days after the beginning of the universe, therefore confirming the Biblical account of creation that revealed about 3500 years ago that the formation of the Earth was completed on the 3rd day, while that of the Moon and the Sun was completed on the 4th day of creation. Nathanael-Israel Israel is referred to as the "Undisputable Specialist of all Questions at the Intersection of Science and Biblical Creation."

FROM SCIENCE TO BIBLE'S CONCLUSIONS
THE #1 UNIVERSE-ORIGIN MASTERPIECE OF ALL TIME … AND THE MOST ACCURATE SCIENTIFIC FORMULA THAT STOOD AND WILL STAND THE TEST OF TIME AND OF MATHEMATICS

The real reason scientists have been struggling to accurately understand the universe's formation is because they have spent centuries collecting expensive, complicated, and massive amounts of data but learned very little, if not nothing, about how to unconventionally step back to properly analyze it to decode the universe. Consequently, people learned to collect all kinds of data everywhere to build models and imaginary concepts that betray their discernment, but they never learned to unlearn wrong theories, nor learned how to stop trashing great raw data hidden in theories they dislike or misunderstand, nor never knew where to find and how to properly combine the fundamental variables without which it is impossible to ever clear the way so their data can properly work for and precisely lead them to the real origin of the universe. How can people abandon the dangerous theories they think are correct because they don't know any better ones?

Lucky you, that is where Dr. Nathanael-Israel Israel, the founder of Science180 (Science180.com) came in to properly reanalyze and put under control these costly, underrated data to provide the accurate and simple solution people have been looking for throughout the ages but that they have ignored.

In *"From Science to Bible's Conclusions,"* you will:

- Get a world-class explanation of the 4 fundamental variables without which it is unquestionably impossible to ever decode the universe's formation scientifically

- Save time and money, and enjoy a life filled with the wonderful peace that the accurate understanding of the universe's origin can create

- Discover the errors in the scientific theories and religious belief systems about the universe's formation that are putting you at risk, and learn how to take control over cosmological threats lurking at the edge of your rational mind, faith, disbelief, or doubt

- Unlock the accurate scientific formula to rationally test the existence of God in a historic way that uncompromisingly satisfies both believers and skeptics (Science180.com/public)

- Get all you need to become a knowledgeable person who will never again need anybody else to explain to you the origin of the universe, for, you will fully understand and articulate it yourself and rationally know whether science is really at war with religion
- Receive deep insights that even those who went to university for years were not able to decrypt by themselves, so you can equip yourself to eliminate all forms of scientific and religious universe-origin prejudices
- Discover whether the scientific data finally confirms that the formation of the Earth was completed on the 3rd day, while that of the Moon and the Sun was on the 4th day of creation like the Bible says, or whether the data proves that it took billions of years to progressively form the universe
- Understand the celebrated scientific formula that rationally puts to rest all debates about the relationship between science, faith, and all theories about the universe's origin so you can properly develop yourself, expand your network, and shape your future

Quickly grab and read this scientifically verifiable, bestselling book to finally get the accurate, jaw-dropping answer that has been rationally shaking believers, skeptics, and all freethinkers. Don't wait!

Dr. Nathanael-Israel Israel has had the honor to be acknowledged as the #1 universe-origin, life-origin, and chemicals-origin expert. He is the author of *"Turbulent Origin of the Universe"*, *"Reconciling Science and Creation Accurately"*, *"Turbulent Origin of Chemical Particles"*, *"Turbulent Origin of Life"*, *"How Baby Universe Was Born"*, *"Science180 Accurate Scientific Proof of God"*. Visit Israel120.com to learn more about this world's most trusted expert that helps scientists and laypeople to properly decode the origin and formation of the universe, life, and chemicals so people can live more effectively nonstop.

TURBULENT ORIGIN OF LIFE

THE ONLY ACCURATE FORMULA TO SCIENTIFICALLY EXPLAIN THE FORMATION OF ALL FORMS OF LIFE QUICKLY

Every human being will benefit from understanding the real origin of life. But the problem is that most efforts to explain the origin of life are complex, inaccurate, confusing, partisan, and complicated, therefore creating an accurate, simple, straightforward, nonpartisan life-origin book that is free from jargon and difficult concepts only known by the experts. This elegant scientific book breaks down the technicality of the origin of life in a language that even nonscientists can easily comprehend. It is a trustworthy book that will help you to quickly, cheaply, easily, and efficiently navigate everything you need to know to finally decode and solve the puzzling problems about the origin of life, while also giving you a crash course on the universe's origin.

Unlike any book you have ever read on the origin of life, this historic masterpiece (that distills complex scientific data down to simple explanations that make sense) is the starting point for any smart person wanting to rationally understand the formation of all living things. By the time you finish reading *"Turbulent Origin of Life,"* you will discover:

- Why, in spite of the massive amount of scientific data collected on living things, scientists have misunderstood the formation of life until now, and then uncover in a simple language the one thing that was needed to accurately crack the code of life but that scientists have missed and that has been causing them headaches, overwhelm, and burnout

- Step-by-step pathway to decode the origin of life and get the power, freedom, and boldness to take advantage of the opportunities that the accurate understanding of the origin of life creates (Science180.com/life)

- The high connection between the code of the universe formation and the process by which life on Earth was formed so you can become a fulfilled thought leader in your field of expertise

- Tools to stand as a lightning bolt that electrifies those who are still struggling to understand the formation of all forms of life in the universe

- Strategies to push the boundaries of human abilities to properly understand what is perceived as un-understandable, mysterious, supernatural, unimaginable, impossible, and unthinkable that hold people back

- Scientific approach to holistically detect, correct, and remove all misinformation, ambiguity, and misleading claims and theories surrounding the origin of life

Whether you are a scientist or a layperson, a believer or a skeptic, you cannot afford to ignore the greater, better, faster, simpler, cheaper, easier, and more accurate formula unlocked in this important book that successfully decoded the origin of life. Get *"Turbulence Origin of Life"* today and change lives! Don't wait!

Dr. Nathanael-Israel Israel is the father of Science180 Cosmology and the founder of Science180 Academy. He is fortunate to be known as the source of unconventional wisdom and knowledge that helps people accurately crack the code of the formation of the universe, of life, and of chemicals. Get some resources by visiting his personal website at Israel120.com.

HOW GOD CREATED BABY UNIVERSE
THE FIRST AND ONLY BOOK THAT ACCURATELY EXPLAINS EVERYTHING ABOUT THE FORMATION OF THE UNIVERSE AND LIFE IN A WONDERFUL LANGUAGE THAT ALL CHILDREN AGES 7-12 CAN EASILY AND FULLY UNDERSTAND & ENJOY!

As the only universe-origin book that your whole family will like and enjoy together, *"How God Created Baby Universe"* will set children on the path of success by accurately showing them early in life the formation of the universe and how to detect errors in theories or stories that would misguide them as they grow up. Therefore, you need to add this great, efficient, trustworthy, and cost-effective book to the strategic journey of children toward their best tomorrow.

With *"How God Created Baby Universe,"* you will:

- Have a peace of mind that children will get accurate, and easy-to-understand universe-origin information that will produce real results in their lives.

- Become the leader that captures the heart of children craving for the original explanation of the formation of the universe so you can clear their way for freedom, power, technology, innovation, and breakthroughs of the future (learn more at Science180.com/children)

- Protect yourself and loved ones from wrong theories in the literature and the media by keeping children secured and empowered with the true knowledge of how the universe began

- Explain complicated secrets to children about how to locate mistakes in origin-related theories so you can save time, money, and other resources to improve their lives

- Ultimately boost children's confidence in detecting, confronting, and avoiding wrong theories by knowing the facts and real processes involved in the formation of the universe

- Help children to easily sort out their origin-related questions using strategies that get them to tap into deep secrets that even highly educated people ignore

- Clearly explain to children how to mathematically know if God created the universe as the Bible says or billions of years evolution processes formed it

Accurately explaining the complex formation of the universe and of life to children can be very hard in our modern world, but by getting *"How God Created Baby Universe"*, you will know the proven formula to help children to easily understand their huge universe-origin and life-origin questions with confidence, humor, and joy. They will surely laugh aloud while reading this book and thank you for it! It is time to buy this pragmatic book to help the children in your life today.

A member of the American Association for the Advancement of Science, the American Chemical Society, and the American Society for Microbiology, **Dr. Nathanael-Israel Israel is** a Beninese-American scientist and international consultant who shows the world how to scientifically decode the formation of the universe, of life and who is known as the creator of the Chemicals Turbulent Origin Formula™, the inventor of the Life Turbulent Origin Formula™, and the discoverer of the Universe Creation Formula™. Learn more at Israel120.com.

NATHANAEL-ISRAEL ISRAEL: WHO IS TOLD BY PEOPLE THAT HE IS THE
UNIVERSE-ORIGIN, LIFE-ORIGIN & CHEMICALS-ORIGIN ACCURATE
DECODER

NOTES

Chapter 1. Can you scientifically prove the existence of God to atheists and all freethinkers without making the believers unhappy?

1. Holden Michael (2010). God did not create the universe, gravity did. http://blogs.reuters.com/faithworld/2010/09/02/god-did-not-create-the-universe-gravity-did-says-stephen-hawking/. Visited on December 18, 2013.

2. Lipka Michael, Tevington Patricia, Starr Kelsey (7 February 2024). "8 facts about atheists". Pew Research Center.

3. Wikipedia (2025). Atheism in the United States. https://en.wikipedia.org/wiki/Atheism_in_the_United_States, Visited May 7, 2025.

4. PRRI (2024). Religious change in America. PRRI. 27 March 2024.

Chapter 2. How to impartially investigate God's existence pains and expectations in the world to properly handle faith and doubt scientifically

5. Rational responders (2024). How to irritate an atheist. www.rationalresponders.com/how_to_irritate_an_atheist, Visited April 4, 2024.

6. Randall Hardy (2024). https://amen.org.uk. Visited April 2, 2024.

Chapter 4. Historic story of the remarkable "Universe Turbulent Origin Formula" that everybody is talking about

7. NASA (2018). Planetary fact sheets. Fact sheets of the Sun, planets, satellites, rings, and selected asteroids in the Solar System. Author/Curator: Dr. David R. Williams, NASA Goddard Space Flight Center, Greenbelt, MD, USA. Retrieved on November 19, 2018, from http://nssdc.gsfc.nasa.gov/planetary/factsheet/.

Chapter 5. Can any creation myth or religious story in the world scientifically teach us anything about God? If not, why?

8. Leonard Scott A; McClure, Michael (2004). Myth and Knowing (illustrated ed.). McGraw-Hill. ISBN 978-0-7674-1957-4.

9. Bhikku Bodhi (2007). "III.1, III.2, III.5". In Access To Insight (http://www.accesstoinsight.org/tipitaka/dn/dn.01.0.bodh.html). The All Embracing Net of Views: Brahmajala Sutta. Kandy, Sri Lanka: Buddhist Publication Society.

10. Georgis Faris (2010). Alone in Unity: Torments of an Iraqi God-Seeker in North America. Dorrance Publishing. p. 62. ISBN 1-4349-0951-4.

11. Torrance Robert M. (1999). Encompassing Nature: A Sourcebook. Counterpoint Press. pp. 121–122. ISBN 978-1-58243-009-6. Retrieved 15 December 2012.

12. Kenneth Kramer (1986), World scriptures: an introduction to comparative religions, p. 34, ISBN 978-0-8091-2781-8.

13. McNamara Patrick and Wesley J. Wildman (2012). Science and the World's Religions [3 Volumes]. ABC-CLIO. pp. 180–. ISBN 978-0-313-38732-6. 19 July 2012. Retrieved 15 December 2012.

14. Numbers Ronald L. (2006) [Originally published 1992 as The Creationists: The Evolution of Scientific Creationism; New York: Alfred A. Knopf]. The Creationists: From Scientific Creationism to Intelligent Design (Expanded ed., 1st Harvard University Press pbk. ed.). Cambridge, MA: Harvard University Press. ISBN 0-674-02339-0. LCCN 2006043675.OCLC 69734583.

15. Wikipedia (2023). Creationism. Retrieved on February 24, 2023, from https://en.wikipedia.org/wiki/Creationism.

16. Islam for Today (2007). The Origin of Life: An Islamic perspective". Retrieved 2007-03-14. http://www.islamfortoday.com/emerick16.htm.

17. Gallup (2013). Evolution, Creationism, Intelligent Design. Retrieved 4 May 2013. http://www.gallup.com/poll/21814/Evolution-Creationism-Intelligent-Design.aspx.

Chapter 6. Do you have to embrace evolution or deny Biblical creation to scientifically prove that God created the universe in 6 literal days?

18. Clarke Adam (1832). Commentary on Genesis 1. The Adam Clarke Commentary. Retrieved on October 5, 2017, from www.studylight.org/commentaries/acc/genesis-1.html.)

Chapter 9. Why am I teaching the whole world how science accurately supports faith and a bizarre creation story ... and what can you do to avoid the dangerous thing that will happen next?

19. Israel Nathanael-Israel (2025a). Turbulent Origin of the Universe. Science180, Augusta, USA 683 pages.

20. Israel Nathanael-Israel (2025b). From Science to Bible's Conclusions. Science180, Augusta, USA 170 pages.

21. Israel Nathanael-Israel (2025c). Reconciling Science and Creation Accurately. Science180, Augusta, USA 299 pages.

22. Israel Nathanael-Israel (2025d). Turbulent Origin of Chemical Particles. Science180, Augusta, USA 397 pages.

23. Israel Nathanael-Israel (2025e). Turbulent Origin of Life. Science180, Augusta, USA 370 pages.

24. Israel Nathanael-Israel (2025f). Origin of the Spiritual World. Science180, Augusta, USA 151 pages.

25. Israel Nathanael-Israel (2025g). How Baby Universe Was Born. Science180, Augusta, USA 130 pages.

26. Israel Nathanael-Israel (2025h). How God Created Baby Universe. Science180, Augusta, USA 224 pages.

27. Israel Nathanael-Israel (2025i). Science180 Accurate Scientific Proof of God. Science180, Augusta, USA 214 pages.

28. Israel Nathanael-Israel (2026a). Mathematical Proof of God's Existence at the Intersection of Science and Faith. Science180, Augusta, USA 74 pages.

INDEX

SCIENCE180: THEORY THAT HELPS YOU AVOID DANGEROUS DOGMA AND IRRATIONAL THINKING

SCIENCE180: THEORY THAT HELPS YOU AVOID DANGEROUS DOGMA AND
IRRATIONAL THINKING

SCIENCE180: THEORY THAT HELPS YOU AVOID DANGEROUS DOGMA AND IRRATIONAL THINKING

SCIENCE180: THEORY THAT HELPS YOU AVOID DANGEROUS DOGMA AND
IRRATIONAL THINKING

ABOUT THE AUTHOR

Dr. Nathanael-Israel Israel scientifically challenged, demystified, and changed the way scientists, nonscientists or laypeople, believers, unbelievers, and all kinds of freethinkers (e.g., atheists, humanists, skeptics, evolutionists, and anti-creationists) think about the formation of the universe, of life, and of chemicals. Using a unique insight that courageously contradicts conventional scientific methods and religious boundaries, Dr. Israel has earned the reputation as the world's #1 authority on complex issues pertaining to the origin of the universe, of life, and of chemical particles. Because no one has ever imagined that a human being can accurately decipher the blueprint of the formation of the universe like Nathanael-Israel Israel did, people of all scientific, religious, freethinking, and political backgrounds have been seeking his historic expertise to unconventionally decode the origin of the universe and improve lives nonstop.

Dr. Nathanael-Israel Israel is a member of the American Chemical Society, American Association for the Advancement of Science, American Society of Agricultural and Biological Engineers, American Society for Microbiology, American Society of Biochemistry and Molecular Biology, Ecological Society of America, American Society of Agronomy, Crop Science Society of America, and Soil Science Society of America. A renowned personality in the universe-origin, life-origin, and chemicals-origin space, Dr. Nathanael-Israel Israel is the founder of Science180, the American organization that operates Science180 Academy (Science180Academy.com), a non-degree training, speaking, consulting, and mentoring program designed to groom and empower people of all backgrounds in the truth about the origin of the universe, life, and chemicals. Before launching Science180, Dr. Nathanael-Israel Israel worked as a scientist at a major Fortune 500 biotechnological  company in the USA, where this Beninese American also founded and owns a news company. Some of his groundbreaking books include:

Turbulent Origin of the Universe

SCIENCE180: THINK GOD'S EXISTENCE DIFFERENTLY & WHERE GOD'S EXISTENCE IS ACCURATELY DECODED, FULL STOP

ABOUT THE AUTHOR

Reconciling Science and Creation Accurately
Turbulent Origin of Chemical Particles
Origin of the Spiritual World
From Science to Bible's Conclusions
Turbulent Origin of Life
How Baby Universe Was Born
How God Created Baby Universe
Science180 Accurate Scientific Proof of God
Mathematical Proof of God's Existence at the Intersection of Science and Faith

Get these thoughtful books to figure out what happened at the beginning of the universe, what is coming up, and why it is time to urgently rethink everything you have been told about the universe's origin, so you don't eventually regret! Connect with this historic scientist and get free resources today by visiting Israel120.com.

NATHANAEL-ISRAEL ISRAEL: DISCOVERER OF THE FORMULA AT THE
INTERSECTION OF SCIENCE AND GOD'S EXISTENCE

www.ingramcontent.com/pod-product-compliance
Lightning Source LLC
Chambersburg PA
CBHW070914130626
46555CB00001B/127